EU Climate Diplomacy

AF845130

The European Union has long played a leadership role in the global response to climate change, including the development and dissemination of climate-friendly technologies such as renewable energy. EU diplomacy has been a vital contributor to the development of international cooperation on climate change through the agreement of the United Nations Climate Convention, its Kyoto Protocol and, most recently, the Paris Agreement. In addition, the election of Donald Trump as President of the United States means that the EU contribution to climate diplomacy will become more important still, both in filling the leadership gap (together with other major economies) and in responding to any sabotage by the Trump administration.

This book will extend knowledge of the EU as a key actor in climate diplomacy by bringing together leading practitioners and researchers in this field to take stock of the EU's current role and emerging issues. Contributions will be grouped into three strands: 1) the interplay between EU climate diplomacy and internal EU politics; 2) how the EU's legal order is a factor that determines, enables and constrains its climate diplomacy; and 3) the EU's contribution to diplomacy concerning climate technology both under the Climate Convention and more broadly. Collectively, these contributions will chart the EU's role at a critical time of transition and uncertainty in the international response to climate change.

EU Climate Diplomacy: Politics, Law and Negotiations will be of great relevance to students, scholars and policymakers with an interest in international climate politics and policy, transnational environmental law and politics and EU studies more generally.

Stephen Minas is an Assistant Professor at the School of Transnational Law, Peking University, China, and Senior Research Fellow in the Transnational Law Institute at King's College London, UK.

Vassilis Ntousas is the International Relations Policy Advisor at the Foundation for European Progressive Studies, Belgium.

Routledge Advances in Climate Change Research

Recently published titles:

Climate Change and Food Security
Africa and the Caribbean
Edited by Elizabeth Thomas-Hope

Globalizing the Climate
COP21 and the Climatisation of Global Debates
Edited by Stefan C. Aykut, Jean Foyer and Edouard Morena

Adapting Infrastructure to Climate Change
Advancing Decision-Making under Conditions of Uncertainty
Todd Schenk

Local Action on Climate Change
Opportunities and Constraints
Edited by Susie Moloney, Hartmut Fuenfgeld and Mikael Granberg

Pricing Carbon in Australia
Contestation, the State and Market Failure
Rebecca Pearse

The Paris Framework for Climate Change Capacity Building
Mizan R. Khan, J. Timmons Roberts, Saleemul Huq and Victoria Hoffmeister

The Anthropology of Climate Change
An Integrated Critical Perspective
Hans A. Baer and Merrill Singer

EU Climate Diplomacy
Politics, Law and Negotiations
Edited by Stephen Minas and Vassilis Ntousas

The Global Climate Regime and Transitional Justice
Sonja Klinsky and Jasmina Brankovic

For more information about this series please visit: www.routledge.com/Routledge-Advances-in-Climate-Change-Research/book-series/RACCR

"Thanks for this great book. Climate politics has become the crucial field to win a future or lose track. A brave new world is being shaped and Europe cannot miss its chance to become a pivotal player. Time has come to dilute our differences and stress European common values and shared perceptions of the challenges ahead of us. Equity, social justice, environmental limits and distributional policies are the key parameters to gain confidence in a shared prosperity. They deserve full political engagement."

– **Teresa Ribera**, *Director, Institute for Sustainable Development and International Relations, France*

"At a time when international climate policy is resetting itself to implement the Paris Agreement, *EU Climate Diplomacy: Politics, Law and Negotiations* is an important contribution that offers valuable insights on the European Union's leadership on its approach to climate change."

– **Martijn Wilder AM**, *Partner, Baker & McKenzie, Australia*

"While the EU's political clout has dwindled, it often led the international climate negotiations by good example. Its strategy to accommodate the US, however, has backfired. The book is, therefore, crucial to understand the EU's current role and also a timely assessment of the way forward."

– **Dr Christoph Schwarte**, *Executive Director, Legal Response Initiative, UK*

"This book is for me a great contribution to the discussions in the foreign policy domain at EU level, as I am introducing climate diplomacy as a topic into the European Parliament. As climate is a crucial Sustainable Development Goal, Europe has to deliver internally and externally and this book greatly complements this reflection process."

– **Arne Lietz**, *Member of the European Parliament for the Group of the Progressive Alliance of Socialists and Democrats, Germany*

EU Climate Diplomacy

Politics, Law and Negotiations

Edited by Stephen Minas
and Vassilis Ntousas

Routledge
Taylor & Francis Group
LONDON AND NEW YORK

earthscan
from Routledge

First published 2018 by Routledge

2 Park Square, Milton Park, Abingdon, Oxfordshire OX14 4RN
52 Vanderbilt Avenue, New York, NY 10017

Routledge is an imprint of the Taylor & Francis Group, an informa business

First issued in paperback 2019

© 2018 selection and editorial matter, Stephen Minas and Vassilis Ntousas; individual chapters, the contributors

The right of Stephen Minas and Vassilis Ntousas to be identified as the authors of the editorial material, and of the authors for their individual chapters, has been asserted in accordance with sections 77 and 78 of the Copyright, Designs and Patents Act 1988.

All rights reserved. No part of this book may be reprinted or reproduced or utilized in any form or by any electronic, mechanical, or other means, now known or hereafter invented, including photocopying and recording, or in any information storage or retrieval system, without permission in writing from the publishers.

Trademark notice: Product or corporate names may be trademarks or registered trademarks, and are used only for identification and explanation without intent to infringe.

British Library Cataloguing-in-Publication Data
A catalogue record for this book is available from the British Library

Library of Congress Cataloging-in-Publication Data
A catalog record for this book has been requested

ISBN: 978-1-138-09728-5 (hbk)
ISBN: 978-0-367-43023-8 (pbk)

Typeset in Goudy
by Florence Production Ltd, Stoodleigh, Devon, UK

To Elpida – VN

To Stacey – SM

Contents

Figures

Tables

Contributors

Eduardo Alvarez Armas obtained his 'Licenciatura en Derecho' (LLB – *Summa Cum Laude*) at the Universidad de La Laguna (Tenerife, Spain) in 2006. Upon graduation, he began working for J&A Garrigues, one of Spain's leading law firms, in their civil and commercial litigation department. In 2008, he moved to Belgium, where he received a Masters in European Law at the College of Europe (2009) and a Masters in International Law at the Université catholique de Louvain (2010 – *Summa Cum Laude*). Between 2010 and 2015, he worked as a research fellow in legal sciences for the Belgian National Fund for Scientific Research (FNRS), teaching and researching at the Université catholique de Louvain. After obtaining an LLM in American Law at Stanford Law School, USA, in 2016, he defended his PhD dissertation in 2017 (double PhD diploma, Université catholique de Louvain, Belgium and Universidad de Granada, Spain – *Summa Cum Laude*). He is currently a Research Associate at Queen Mary University of London, UK.

Emily Barritt is Lecturer in Tort Law at The Dickson Poon School of Law, King's College London, UK, and a member of the King's Climate Law and Governance Hub. Her research focuses on environmental democracy, access to environmental justice, stewardship and climate change adjudication. She was previously a Centre Fellow at the Centre of Environment, Energy and Natural Resource Governance, University of Cambridge, UK, where she worked on a United Nations Environment Programme project developing legal options on marine biodiversity protection for a new implementing agreement to the Law of the Seas Convention. In 2016–2017, she was a lecturer in Law at Corpus Christi College, University of Oxford, UK, teaching on the undergraduate environmental law module.

Katja Biedenkopf is Assistant Professor at the University of Leuven, Belgium. Her research centres on environmental leadership and entrepreneurship, climate diplomacy, policy diffusion, carbon markets, electronic waste and chemicals policy. Previous positions include Assistant Professor at the University of Amsterdam, the Netherlands, Postdoctoral Fellow at the Free University of Berlin, Germany, Fulbright Fellow at the University of California, Berkeley, USA, and Visiting Fellow at Johns Hopkins University, Washington, DC, USA.

Megan Bowman joined King's College London, UK, in 2015 and has been appointed a Dickson Poon School of Law Rising Star. Her research focuses on intersections between commercial, financial and environmental law and regulation in transnational contexts, with particular emphasis on private actors. Her first book, *Banking on Climate Change: How Finance Actors and Transnational Regulatory Regimes are Responding* (Kluwer, 2015), combines empirical and theoretical insights and has been described as a ground-breaking study. Dr Bowman co-founded the King's Climate Law and Governance centre, comprising cutting-edge research, teaching and engagement that focuses on sustainability-related legal and governance challenges and opportunities around the world. She consults to national governments and intergovernmental agencies on implementing the Paris Agreement and advises prominent non-government organizations on sustainable banking and investment issues. She is a qualified barrister and solicitor of the High Court of Australia and Supreme Court of Victoria, with an LLM in International Law and a PhD in regulatory theory.

Stuart Bruce is an Australian-qualified lawyer in the international disputes practice of a major international law firm, focusing on investment treaty arbitration, public international law and international commercial arbitration for corporate and government clients. He has particular interests in sustainable energy, natural resources and environmental matters, both professionally and academically. Stuart is sometime researcher at the University of Cambridge, UK, Legal Research Fellow at the Centre for International Sustainable Development Law, Fellow of the Platform on International Energy Governance, co-Chair of the World Bank's Legal Aspects of Sustainable Energy for All Community of Practice within the Global Forum on Law, Justice and Development and member of the IUCN World Commission on Environmental Law. Stuart holds an LLM (International Law) with Distinction (UCL), a Graduate Diploma in Energy and Natural Resources Law (Melbourne, Australia) and an LLB (Monash, Australia).

George Dellis is Associate Professor of Public Law at the Athens Law Faculty, lawyer at the Athens Bar Association, a graduate of the Universities of Athens, Greece, Paris II-Assas, France (LLM, Dr. iur.), and honoured with the Habilitation à diriger des recherches, former Référendaire at the Court of Justice of the European Union. Teaching and practicing public and European law since 2000, he has published books and articles on European and national environmental and land planning law, administrative law, administrative courts and procedures, regulated industries, competition law and law and economics, in Greek, French and English. He is the author of two books in Greek, related to the subject: *European Environmental Law* (Ant. Sakkoulas, Athens, 1998) and *Public Utilities and the Market: Vols I and II* (Ant. Sakkoulas, Athens, 2008).

Eugenia Giannini is an Attorney at Law, Member of the Athens Bar Association since 1991 as well as a member of the Association des Juristes de Droit Public Compare (Paris) since 1990 and of the Chartered Institute of Arbitrators

(London). She graduated from the Law School of the University of Athens, Greece, and conducted her postgraduate studies on comparative European public law at the Paris I–Pantheon Sorbonne University, France. She is an accredited arbitrator and mediator from the Chartered Institute of Arbitrators (London) and the Harvard Negotiation Institute. Dr Giannini presented her PhD thesis 'The penetration of RES in the energy system of Greece and the Sustainability Principle' before the National Technical University of Athens, Greece (2017).

Laura Hanning Scarborough has worked at a senior political and executive level as a board member, programme director, climate finance expert and the UN negotiator. The head of the Lithuanian delegation 2009–2010 in the UN climate process, and adviser to the government during the EU Presidency in 2013, Laura served as the vice chair of the UN Adaptation Fund Board and the chair of the Projects and Programmes Committee for 3 years. Laura holds Bachelor of Laws degree from Sussex University, UK, and Master of Studies from Cambridge University, UK. She is now working as an associate partner at the Centre of Multilateral Negotiations.

Matthew Kennedy is Head of Strategy and Business in the International Energy Research Centre (IERC). Matt was lead EU Negotiator (Energy Technology) at COP21 UN Framework Convention on Climate Change (UNFCCC) and was a member of the UNFCCC Technology Executive Committee. Matt is the former Chair of the UN's Climate Technology Centre and Network, Chair of the IEA's Renewable Energy Technology Deployment technology collaboration programme and the Chair of the Programme Board of the Renewable Energy and Energy Efficiency Partnership (REEEP). For the Irish Government, Matt was responsible for Energy R&D for Ireland (Horizon 2020) (Energy) and led Energy R&D for the Sustainable Energy Authority of Ireland. Matt holds a PhD in Engineering from Trinity College Dublin, Ireland, and Masters' degrees from NUI Galway, Ireland, and University College Dublin, Ireland.

Karsten Krause is a Policy Officer in the European Commission's Directorate General for Climate Action. In the Directorate for International and Mainstreaming, he works on international technology transfer and innovation policy. As part of the EU's UNFCCC team, he is coordinating the work on technology development and transfer. Since 2014 he is a nominated member of the UN's Climate Technology Centre and Network's Advisory Board and the EU's National Designated Entity.

Robert Ladrech is Professor of European Politics at Keele University, UK. His research focuses on the relationship between party politics and European integration, with a particular interest in social democratic parties and climate policy. He is also interested in the impact of the EU on domestic politics in general (Europeanization). He recently completed an ESRC project on 'Climate Policy and Political Parties' analysing the positions that mainstream parties of the centre-left and centre-right have taken on climate policy since the mid-1990s. Professor Ladrech has also been a Visiting Professor at the College of Europe, Bruges, Belgium, and is the author of *Europeanization and*

National Politics (Palgrave, 2010) and *Social Democracy and the Challenge of European Union* (Lynne Rienner, 2000), as well as numerous scholarly articles.

Rafael Leal-Arcas, Jean Monnet Chaired Professor in EU International Economic Law and Professor of Law, Queen Mary University of London, UK (Centre for Commercial Law Studies); Member, Madrid Bar. PhD, European University Institute, Italy; JSM, Stanford Law School, USA; LLM, Columbia Law School, USA; MPhil, London School of Economics and Political Science, UK; JD, BA, Granada University, Spain. The financial help from two EU grants is greatly acknowledged: Jean Monnet Chair in EU International Economic Law (number 575061-EPP-1–2016–1-UK-EPPJMO-CHAIR) and the WiseGRID project (number 731205), funded by Horizon 2020. Author of the following books: *Energy Security, Trade and the EU: Regional and International Perspectives* (Edward Elgar, 2016); *The European Energy Union: The Quest for Secure, Affordable and Sustainable Energy* (Claeys & Casteels Publishing, 2016); *International Energy Governance: Selected Legal Issues* (Edward Elgar, 2014); *Climate Change and International Trade* (Edward Elgar, 2013); *International Trade and Investment Law: Multilateral, Regional and Bilateral Governance* (Edward Elgar, 2010) and *Theory and Practice of EC External Trade Law and Policy* (Cameron May, 2008).

Stephen Minas, BA(Hons), LLB(Hons) (Melbourne, Australia), MSc (LSE, UK), GDLP, is Assistant Professor of Law at the School of Transnational Law, Peking University, China. He is a Senior Research Fellow of the Transnational Law Institute and Adjunct Fellow in Climate Law and Governance, both at King's College London, UK, and a member of the UNFCCC Technology Executive Committee's task forces on climate technology financing and mitigation, the IUCN World Commission on Environmental Law and the Young Academics Network of the Foundation for European Progressive Studies (FEPS). He has consulted to or collaborated with a variety of international organizations and think-tanks. Previously, Stephen worked as a visiting lecturer at King's College London, UK (teaching in areas including finance law, climate change law and transnational law), and an adviser to the Premier of the Australian State of Victoria.

Vassilis Ntousas is International Relations Policy Advisor at the Foundation for European Progressive Studies (FEPS), where he coordinates various international projects and activities and conducts political research. Vassilis received his MSc in International Relations from the London School of Economics, UK in 2011, specializing mainly in foreign policy analysis and international security issues. He had previously received a First Class Honours for his BA in International and Politics from the University of Sheffield, UK, in 2008. Prior to joining FEPS, Vassilis worked as senior communications and political advisor at the Municipality of Thessaloniki, Greece, providing advice in the areas of international affairs and intercity diplomatic relations. He has extensive experience in strategic and tactical political communications, with a particular focus on crafting comprehensive, tailor-made messaging, engagement and outreach strategies.

Ernst Stetter is the Secretary General of Foundation for European Progressive Studies (FEPS) since 2008. He is also a regular commentator on EU affairs in the media and visiting fellow at University of Greenwich, London, UK. He is an economist and political scientist. He studied in Tübingen and Heidelberg (Germany), focusing on international trade, finance, economic and social policy as well as development issues. In 1980, he obtained his PhD in political science entitled 'The Association of ACP-Countries (Lomé I and II) to the European Community and the STABEX-System'. In 1976, Ernst began his professional career as a lecturer in economics at the DGB Trade Union Centre for Vocational Training in Heidelberg, Germany. From 1980 to 2008, he worked for the Friedrich Ebert Stiftung (FES) in various positions. He spent the first 4 years at the FES as a consultant in Dakar, Senegal. In 1988, Ernst was appointed as Head of the Africa Department. In 1994, he started working as head of the Central Europe Unit. In 1997, he moved to Paris and became Director of the FES Office in France, and in 2003 he was appointed Director of the EU-Office of Friedrich-Ebert-Stiftung (FES) in Brussels, Belgium. In 2003 he received the French decoration of Chevalier de l'Ordre national du Mérite.

Hayley Walker is a research fellow of the Fonds de la Recherche Scientifique – FNRS at the Université catholique de Louvain, Belgium. Her PhD thesis focuses on the management of multilateral negotiations by chairs and presidencies. Other areas of interest include climate diplomacy and EU/US climate policy and cooperation. She has also been involved in research on capacity building in emissions trading schemes in China. Hayley completed her undergraduate studies at the University of Oxford, UK, and holds a Master's degree in European Studies.

Peer Zumbansen was born in Berlin and educated in Germany, France and the United States. Admitted to the bar in Germany, he studied philosophy and law in Germany and France before receiving an LLM from Harvard Law School, USA, followed by a doctorate and the postdoctoral Habilitation from Frankfurt's Goethe University, Germany. Professor Zumbansen's doctoral thesis won the Walter Kolb Memorial Prize for Best Doctorate in Law. At Osgoode Hall Law School at York University in Toronto, Canada, he held a prestigious Canada Research Chair for a decade – first in the Comparative and Transnational Law of Corporate Governance and, since 2009, in Transnational Economic Governance and Legal Theory. At Osgoode, he was the founding director of the Comparative Research in Law and Political Economy Network (CLPE) and the Collaborative Urban Research Laboratory, two international and interdisciplinary research centres. He was, from 2000 to 2013, the founding co-editor-in-chief of the *German Law Journal* and has been, since 2012, the editor-in-chief of *Transnational Legal Theory*. He founded the CLPE Osgoode Hall Research Paper Series at the Social Science Research Network (SSRN). Professor Zumbansen joined The Dickson Poon School of Law, King's College London, UK in July 2014 as the inaugural Professor of Transnational Law and founding director of the Dickson Poon Transnational Law Institute.

Foreword

Ernst Stetter and Peer Zumbansen

Climate change is one of the defining challenges of our time. The prevention of dangerous climate change and adaptation to the unavoidable effects of climate change are inseparable from basic questions such as food security, the energy system, decent jobs in a viable economy and the well-being of the environment. With the warnings of consensus scientific opinion growing ever more urgent, climate change represents an extreme test of the international community's ability to cooperate in the face of a common threat. With the burden of climate change's effects falling disproportionately on poorer countries, and on disadvantaged communities within countries, our response to climate change is also a profound question of social justice.

The European Union and its Member States have long been crucial leaders in the response to climate change, through domestic policies and international aid and through multilateral and bilateral diplomacy to raise the sights of collective action. Now, with the United States under President Donald Trump having announced its intention to withdraw from the Paris Agreement and renege on climate finance commitments, the onus on the EU to provide leadership in the diplomacy of climate change has never been greater.

This book responds to the challenge of taking stock of EU climate diplomacy and, in particular, opportunities to strengthen EU contributions to ambitious climate action. The book is the product of a collaboration between the Foundation for European Progressive Studies (FEPS), the Transnational Law Institute of King's College London and Fondation Jean-Jaurès. As part of this collaboration, FEPS hosted a policymakers' workshop in Brussels in June 2016 on EU climate diplomacy, with participation from the Party of European Socialists, the European Commission, Member State officials, the Secretariat of UN Framework Convention on Climate Change, civil society and academia. The workshop was followed by a panel discussion in London in October 2016, at which a comprehensive report on the topic, with a foreword by Swedish Foreign Affairs Minister Margot Wallström, was launched.

The present volume builds on those discussions and also reflects subsequent developments, most notably the US repudiation of the Paris Agreement and the current drive to reinvigorate the European project with twenty-seven Member States. The chapters of this book highlight the multifaceted nature of EU climate diplomacy, encompassing specialized processes of technology and finance and ranging from inter-state negotiation to the mobilization of networks of non-state actors. The EU's domestic legal order and policy coherence are also highlighted as key assets for projecting influence in favour of stronger international action on climate change.

Although climate diplomacy is often the domain of technical experts, it is fundamentally inseparable from politics. In particular, progressive politics has been a crucial enabler of effective EU climate diplomacy, through the crafting of ambitious national policy, the deliberations of the Council of the EU and the forging of 'high-ambition' alliances with social partners and with the states and peoples of the Global South. The direction of EU climate diplomacy is therefore interlinked with the quest for a more social Europe, a more democratic EU and, more specifically, a just transition to the low-emissions economy which respects labour and human rights.

Bringing together the insights of participants in climate diplomacy and leading scholars, *EU Climate Diplomacy* intervenes in current debates over the nature of contemporary diplomacy, the collective response to climate change and the direction of the EU. We trust that *EU Climate Diplomacy* will make a positive contribution to both the analysis of key diplomatic processes and the advancement of ambitious and progressive EU leadership on this critical issue.

Acknowledgements

This book has its origins in a 2016 collaboration on the topic of EU climate diplomacy between the Foundation for European Progressive Studies (FEPS), the Transnational Law Institute (TLI) of King's College London and Fondation Jean-Jaurès. As such, there are many contributors whom we gratefully acknowledge, beginning with the authors who have contributed to this book, who have drawn on diverse experiences as participants in, and scholars of, climate politics, law and diplomacy. It has been our privilege to help assemble their collective wisdom and insights.

From FEPS, Charlotte Billingham was instrumental in bringing together project stakeholders, coordinating the project within FEPS' broader activities on climate change and energy and grounding sometimes technocratic discussions in an important political focus on a socially just transition. From Fondation Jean-Jaurès, we are grateful to Blaise Gonda (now of Agence française de développement) for his insightful contributions to the design of the project and also for his participation as a workshop speaker.

The theme and focus of this book were shaped by a policymaker's workshop held in Brussels, a panel discussion in London and a report on the topic of EU climate diplomacy, all in 2016. We are grateful to FEPS for hosting the Brussels workshop and to each of the speakers who contributed their insights: Dr Emily Barritt of King's College London; Charlotte Billingham; Dr Megan Bowman of King's College London; Seb Dance MEP; Blaise Gonda; Dr Matt Kennedy of Ireland's International Energy Research Centre; Sanjeev Kumar of the Change Partnership; Professor Robert Ladrech of Keele University; Professor Louis Lemkow of the Universitat Autònoma de Barcelona; Dr Christian Linder of the European Commission; Stephen Minas; Matthew Phillips, formerly of the Office of the Executive Secretary of the UN Framework Convention on Climate Change (UNFCCC); Yonnec Polet, Deputy Secretary General of the Party of European Socialists; Michael Rantil of the Swedish Energy Agency; Kirsty Schneeberger, now of ClientEarth; Dr Ernst Stetter of FEPS; Stephanie Venuti, now of Baker McKenzie, and Lidia Wojtal, then of Poland's Ministry of Foreign Affairs.

We are also grateful to those who participated remotely: Dr Sander Chan of the German Development Institute; Asher Lessels from the UNFCCC Secretariat; Jyoti Prasad Painuly from the Copenhagen Centre on Energy Efficiency and Paul Watkinson, principal negotiator on climate change for the Government of France.

We acknowledge the contributors to the study on 'The future of EU climate change technology and sustainable energy diplomacy': Margot Wallström, the Minister of Foreign Affairs of Sweden; Miriam Dalli MEP and Stephen Minas. This report was launched at a panel discussion in London. We are grateful to King's College London for hosting the discussion, as well as to the speakers Barry Gardiner MP, Nick Mabey of E3G, Francesca Manchi of the European Commission Representation in the United Kingdom and Dr Yuliya Yurchenko of the University of Greenwich.

Stephen Minas is grateful to Professor Peer Zumbansen, Director of the TLI, for his wonderful support and encouragement generally and for supporting this particular collaborative project from the beginning. The TLI was and remains something of a start-up, and under Peer's leadership it has facilitated transnational legal scholarship and innovative collaborative efforts that engage with the great real-world challenges of our time. Helen Bhandari, as Coordinator of the TLI during the period of this project, provided invaluable support. Professor Eloise Scotford, now of University College London, has been a vital influence in developing thinking on climate law and its components, contexts and interactions. The students of the Global Law of Climate Change LLM course at King's College London, through their diverse perspectives and insights, also informed this editor's understanding of the still-developing Paris regime.

Vassilis Ntousas wishes to acknowledge the vital support of Dr Ernst Stetter, Secretary General of FEPS, towards this project and this publication. It is extremely rare to find an environment as intellectually enriching and professionally empowering as FEPS, and for this, there is a tremendous amount of appreciation. A great deal of gratitude is also reserved for a number of colleagues and friends, notably Giorgos Georgalas and Maria Freitas, for their continuous encouragement, analytical advice and invaluable assistance throughout this beautiful journey.

We also gratefully acknowledge the team at Routledge, who commissioned this book and then expertly guided it to completion: Margaret Farrelly, Hannah Ferguson, Annabelle Harris and Matthew Shobbrook.

Stephen Minas and Vassilis Ntousas

Section 1

Introductory material

1 Introduction

EU climate diplomacy in a time of disruption

Stephen Minas and Vassilis Ntousas

Climate change is the ultimate collective action problem facing the international community, with broadly distributed culpabilities and consequences that incentivize states to free-ride on the efforts of others. As such, even in the 'best of all possible worlds', it would be extremely difficult to solve. In our world, states, the private sector and civil society face the climate challenge in far less serene circumstances, not least including wilful abdication of responsibility and leadership by the government of the United States. The European Union, which has long been a leader in the crafting of collective responses to climate change, now bears an even greater share of the task of charting a progressive way forward.

Disruption is often presented as an unalloyed positive, and indeed, it is clear that technological disruption is making possible more ambitious responses to climate change. Already, renewable energy has played a major role in decoupling energy-related carbon dioxide emissions from global economic growth (International Energy Agency 2016). Technological innovation in areas such as electricity generation, digitalization of energy systems and improved storage capacity all promise to further disrupt the economics of energy in favour of greenhouse gas emission reductions (Helm 2017, pp. 62–88).

And yet, springing from the same dynamics of globalization and digitalization have also come the disruption of work, creating a new 'precariat' of zero-hours contractors (Standing 2014) and robbing millions of the dignity of labour, and the disruption of collective bonds of identity and solidarity, giving rise to what Pope Francis has labelled the 'globalization of indifference' (Friedman 2016). In turn, these economic disruptions have fed political disruption, stoked for years by distributive injustice and financial crisis and erupting in developed country elections over the past few years.

Multilateral diplomacy has also been disrupted as a result, upended by a wave of 'populist' politics. The United States has turned against multilateralism under a reactionary and inward-looking leadership, pledging to withdraw from the Paris Agreement and undermining cooperation in fora such as the United Nations, the G7 and G20. The United Kingdom has voted to leave the European Union, creating a discretionary crisis and forcing the EU institutions and Member States to devote resources and attention to the resulting 'divorce' proceedings. In short, ambitious projects for 'global governance' (Lamy 2012) and the provision of 'global public goods' (Kaul, Grunberg and Stern 1999) are checked by competing, centrifugal forces of alienation, disaggregation and separatism at various levels – of certain countries from the international community, of the Member State from the EU, of the province from the state and of the individual from society.

Finally, there is the biggest disruption of all to contend with: the disruption of natural climate systems through anthropogenic (human-induced) interference, leading to global warming, ocean acidification and future consequences of ultimately unknowable severity (Intergovernmental Panel on Climate Change 2014). The breaching of 'planetary boundaries' through climate change, biodiversity depletion and other planetary-level risks is pushing natural systems beyond the 'safe operating spaces' in which humanity has thrived (Rockstrom et al. 2009). Such has been the impact of human economic activity on the environment that some scientists have argued that human activity has resulted in the earth entering a new geological epoch: the Anthropocene (Galaz 2014, p. 2).

These problems have long been acknowledged, and since 1994, the UN Framework Convention on Climate Change (UNFCCC) has bound nations to the goal of preventing dangerous human impact on the climate system. But with each passing year, the window of opportunity for achieving the Convention's objectives narrows. The Paris Agreement of 2015 was a major diplomatic breakthrough, but the challenges of implementation are immense.

It is in this context that an assessment of the European Union's present – and potential – contributions to the diplomacy of climate change has become necessary. With the increasingly competitive economics of clean energy technologies and the entry into force of the Paris Agreement, there is significant momentum for building upon EU market and normative leadership. With the announced defection of the United States from the international climate regime, EU leadership has never been more important and indeed necessary.

This book analyses both the internal and the external dimensions of the EU's engagement in the diplomacy of climate change. It examines how the domestic political and legal orders of the EU and its Member States shape, enable and constrain its international negotiating positions. It goes on to consider the EU contribution to a particularly important area of climate diplomacy, that of technology and sustainable energy, which is pursued through multiple for a including the UNFCCC and its constituted bodies, the framework of the Sustainable Development Goals, trade negotiations and overseas development assistance. The chapters that follow reveal the EU to be uniquely capable of pushing climate action forward, through the combination of its normative, legal, commercial, diplomatic and overseas development capabilities. With its network of institutions and Member States, the EU is also uniquely placed to contribute to the multiple international processes that address climate change. However, the chapters in this volume also reveal that the EU is challenged to shape and respond to emerging developments, such as the new approaches to technology under the Paris Agreement and the promotion of sustainable technology in a new generation of trade agreements.

Politics

In the multi-level game of EU decision-making, climate diplomacy is unavoidably political. Negotiations between Member States in the Council, with important inputs from the Commission and Parliament, result in the adoption of EU mandates to carry into international negotiations. In Chapter 2, Robert Ladrech argues that the party politics of Member State governments is an important determinant of the level of ambition of EU climate policy, consequently shaping the possibilities of EU climate diplomacy. Ladrech finds that the party leading a coalition government has usually been the most influential in shaping its climate policy and that centre-left parties have generally adopted more ambitious climate positions than centre-right parties.

The Paris Agreement may itself bear the stamp of party composition. The Paris Agreement was in significant part a triumph of EU and Member State diplomacy, through the assembling of the 'High Ambition Coalition' and, in particular, through the French conference presidency's adroit conduct of both pre-conference diplomacy and the Paris climate conference itself. Through an 'eminently political process' (Ourbak 2017, p. 9), the top leadership of the Hollande administration allocated responsibilities for the pre-conference diplomacy and engaged in a global campaign to influence countries and non-state actors to support an ambitious Paris outcome. At the conference itself, the French presidency fostered party ownership of the process through an inclusive, open approach, respecting the rules of negotiation. Grounded in a progressive administration, the French presidency was also able to play the negotiation 'to the left' (Gonda 2016, p. 61), building vital bridges with the developing country caucus of the G77 and China.

However, returning to our theme of disruption, progressive politics, too, is changing, in response to the multiple crises that have menaced Europe and in response, also, to the electoral challenge of extreme right parties and even certain conservative parties that have proved eager to ape their tactics and positions. In Germany, the long period of 'grand coalitions' of the principal centrist parties is subject to growing pressures. In France, the experiment of La République En Marche offers a new departure for social liberalism – with the destination unknown.

Following the political upheavals of the last 2 years, there is now the opportunity to build coherence across the twenty-seven Member States around a set of shared values and normative goals, prominently including climate action. There are promising early signs. In October 2017, the European Parliament stressed that 'Europe must now take the lead in defending the Paris Agreement in order to secure the future of both our environment and our industries' (European Parliament 2017, par. 10). Also at the European Parliament level, the possibility of transnational lists of candidates may be the next step in nurturing the European polity that, for our generation, is often simply assumed.

Such moves to build direct democracy at the European level would be complemented by the strengthening of 'environmental democracy' through the UNFCCC process. In Chapter 3, Emily Barritt sets out opportunities to increase public participation in the UNFCCC process, building on the success of the Aarhus Convention within Europe and the structured engagement of non-Party stakeholders before and after the Paris conference. Barritt contends that 'climate diplomacy needs to be more democratic'. The EU, with its well-established internal processes to manage and represent diverse interests and multiple stakeholders, is well placed to contribute to this in practice.

Of course, the degree of climate leadership that the EU can provide is shaped not only by its internal dynamics but also by what other actors do on a diplomatic terrain which has fluctuated wildly in the years since the Climate Convention was adopted. Hayley Walker and Katja Biedenkopf demonstrate in Chapter 4 that the EU's 'cognitive leadership' in the early years of climate diplomacy was augmented by 'exemplary leadership' and 'entrepreneurial leadership' during the early 2000s, with the agreement on operationalizing the Kyoto Protocol, including through flexible mechanisms, and the EU's successful discharge of its emission reduction obligations under the first commitment period. However, EU influence was seen to wane in the years around the controversial Copenhagen conference, with EU ambitions not matched by other parties, before recovering in the pre-Paris diplomacy. Now, with the rejection of the Paris Agreement by the United States, the EU faces several potential scenarios, with a strengthened EU–China partnership the most ambitious alternative for steering the negotiations.

Law

The EU's unique structure as a regional integration project built through law has had a significant impact on its climate diplomacy. At the international level, the credibility of EU positions ultimately depends on the EU's ability to implement climate commitments through Union and Member State law. The process of legal implementation is complex, often slow and can result in suboptimal policy outcomes. In Chapter 5, George Dellis and Eugenia Giannini examine how the EU's 2020 renewable energy target has been implemented through the law of one Member State, Greece. Dellis and Giannini argue that domestic regulatory incentives to promote renewable energy supply have been inefficient. The embedding into law of the EU's 2030 climate and energy targets, as advanced in the EU's Nationally Determined Contribution under the Paris Agreement, offers an opportunity to improve the legal frameworks of climate policy implementation, notably through the governance mechanism of the Energy Union.

At the Union level, the EU has a well-elaborated set of laws to govern climate change and sustainable energy, including with respect to new technologies, although these require updating to reflect the 2030 targets and the new Paris architecture. In Chapter 6, Stuart Bruce analyses both the key aspects of EU sustainable energy law and its interaction within the 'decentralized' governance of sustainable energy at the international level. Key international instruments beyond the UNFCCC include the UN Convention on the Law of the Sea and regional seas conventions, regarding the sourcing and production of energy from the sea, and the Energy Charter Treaty and its Protocol on Energy Efficiency, regarding the international trade in energy.

The EU's unique legal order also presents opportunities to add force to its climate diplomacy. In Chapter 7, Stephen Minas examines the Energy Community as a case study of the EU extending its diplomatic reach through its legal order. Under the Energy Community Treaty, non-EU Member States, principally in Eastern Europe, commit to implementing EU energy law in their domestic legal systems. By expanding the *acquis* which must be transposed to include new climate and renewable energy obligations, the Energy Community process can harmonize the regulation of its Contracting Parties with that of the EU, potentially also leading to convergence of positions in the UNFCCC negotiations. Notably, this process relies on the attractiveness of attaining membership of the EU.

Similarly, Megan Bowman demonstrates in Chapter 8 that the financial relations between the EU and external parties also have significant potential to augment climate diplomacy. Bowman argues that the Investment Plan for Europe is an 'untapped opportunity' for the large-scale mobilization of climate finance, including from the private sector. The European Fund for Strategic Investment is the key vehicle for this, as well as the announced expansion of the scope of financeable projects from the EU to also encompass Africa and the EU Neighbourhood. Financing will remain a central issue in climate diplomacy for the foreseeable future. As the United States breaches faith with the world's climate-vulnerable communities, including through reneging on $2 billion in unpaid contributions that the Obama Administration had pledged to the Green Climate Fund, it is the EU that can do most to save the developed world's honour.

Negotiations

At the international level, the practical business of climate diplomacy has long centred on the UNFCCC, with its Conference of Parties constituting the most important forum for negotiation. Alongside finance and capacity-building, technology is a crucial means of implementing the Convention and now the Paris Agreement, particularly through supporting the efforts of developing parties. In the climate context, technology refers to processes and know-how as well as physical infrastructure and products. Regarding climate technologies, the negotiations have in recent years been quite productive, with the creation of new institutions to support national efforts.

In Chapter 9, Matthew Kennedy discusses the emergence of climate technology cooperation under the UNFCCC, particularly through the creation of the Technology Mechanism and its institutions, the Technology Executive Committee and the Climate Technology Centre and Network. Kennedy demonstrates that to maximize climate technology outcomes, coherence of action is required not just among these institutions and UNFCCC Parties but also with important technology initiatives beyond the scope of the UNFCCC, such as the International Energy Agency's Technology Collaboration Programme, the International Renewable Energy Agency and Mission Innovation.

The treatment of technology under the Convention and Paris Agreement continues to develop, reflecting both changing Party circumstances and the gallop of technological progress. In Chapter 10, Karsten Krause illustrates this continued development by reflecting on contemporary efforts to address technological innovation and transformational change through the UNFCCC processes. The EU has been a key player in driving this diplomacy, moving from a linear vision of technological transfer from developed to developing countries to a more holistic approach, addressing all stages of the technology cycle.

Climate diplomacy under the UNFCCC includes also the crafting and oversight of specialized bodies under the Convention. One such body is the Adaptation Fund (AF), which mobilizes finance to support climate adaptation in developing countries. In Chapter 11, Laura Hanning Scarborough considers the complex set of considerations that confront those serving on the board of bodies such as AF, which is an important role of EU climate diplomats.

Importantly, climate diplomacy is not limited to the UNFCCC, and there are multiple opportunities in other fora for creative diplomacy to advance the global response to climate change. In Chapter 12, Rafael Leal-Arcas and Eduardo Alvarez Armas survey the prospects for using trade agreements as a 'novel tool' for addressing climate change. Through a textual analysis of several EU free trade agreements, the authors find an increasing interest in the inclusion of climate-relevant provisions. They propose that the crafting of clauses enabling Parties to protect themselves from the trade consequences of fossil fuel subsidies would be a significant step forward.

In sum, the picture that emerges from this volume is of a European Union that has engaged successfully in the difficult business of global cooperation to address climate change, but which has the capacity to make an even greater contribution in the future. The EU Global Strategy commits the EU to 'lead by example by implementing its commitments on sustainable development and climate change' (European Union 2016, p. 40). This exemplary leadership is no doubt necessary, but a constantly evolving international response to climate change also calls for a cognitive and entrepreneurial leadership that helped to produce, in turn, the Kyoto Protocol and Paris Agreement. Achieving this potential requires concerted effort not just from diplomats and civil servants but also from the politicians, political movements, finance and industry actors, activists and academics who together constitute the grand 'orchestra' of European diplomacy (European Union 2016, p. 46).

References

European Parliament. (2017). Resolution of 4 October 2017 on the 2017 UN Climate Change Conference in Bonn, Germany (COP23) (2017/2620(RSP)).

European Union. (2016). *Shared Vision, Common Action: A Stronger Europe: A Global Strategy for the European Union's Foreign and Security Policy*. https://eeas.europa.eu/archives/docs/top_stories/pdf/eugs_review_web.pdf.

Friedman, U. (2016). "Refugees and the 'globalization of indifference'." *The Atlantic*. www.theatlantic.com/international/archive/2016/04/refugees-pope-francis-lesbos/477870/.

Galaz, V. (2014). *Global Environmental Governance, Technology and Politics: The Anthropocene Gap*. Edward Elgar, London.

Gonda, B. (2016). "Négocier pour la planète. Récits de la COP21." *Les carnets du* CAPS, MAEDI, 45–86.

Helm, D. (2017). *Burn Out: The Endgame for Fossil Fuels*. Yale University Press, New Haven, CT.

Intergovernmental Panel on Climate Change. (2014). Climate Change 2014: Synthesis Report. Contribution of Working Groups I, II and III to the Fifth Assessment Report of the Intergovernmental Panel on Climate Change [Core Writing Team: R.K. Pachauri and L.A. Meyer (eds)]. IPCC, Geneva, Switzerland, 151 pp.

International Energy Agency. (2016). *Decoupling of Global Emissions and Economic Growth Confirmed*. www.iea.org/newsroom/news/2016/march/decoupling-of-global-emissions-and-economic-growth-confirmed.html (accessed October 11 2017).

Kaul, I., Grunberg, I. and Stern, M. (1999). *Global Public Goods: International Cooperation in the 21st Century*. Oxford University Press, Oxford.

Lamy, P. (2012). "Global Governance: From Theory to Practice." *Journal of International Economic Law* 15, 721.

Ourbak, T. (2017). "Analyse retrospective de la COP 21 et de l'Accord de Paris: Un exemple de diplomatie multilatérale exportable?" Rapport d'expertise, MAEDI.

Rockstrom, J., et al. (2009). "A Safe Operating Space for Humanity." *Nature* 461, 24, 472–475.

Standing, G. (2014). *Precariat: The New Dangerous Class* [revised edition]. Bloomsbury Academic, London.

Section 2

Politics

2 Party politics and EU climate policy

Robert Ladrech

An EU climate mitigation policy that has any realistic chance of contributing to fundamental progress with regard to carbon emissions reduction, increasing the proportion of renewable energy and increasing energy efficiency, must, in the end, depend on the agreement and compliance of its member states. The development of the EU's climate policy has been, as with most policy areas involving EU policy competence, a compromise between member state governments whether in terms of targets, deadlines, definition of terms and so on. The Paris Agreement introduced a new or additional variable into this situation, namely, that every 5 years a progress review would take place with an assumption that national ambitions will have been 'ratcheted up' over time. As far as research on EU and member state climate policy is concerned, explanation for a change in policy direction – or more precisely in the level of ambition – is only episodic. In particular, the literature on 'leaders' and 'laggards' does not include analyses of 'leaders' altering their commitment to various aspects of climate policy. Since increased ambition with regard to the Paris Agreement is incumbent upon the EU and its member states, explaining variation in climate policy ambition is important in order for EU-level actors to have an understanding of the motivations of member states when attempting to construct more ambitious policy. This chapter explores one aspect of this phenomenon, that is, the role that political party activity plays in a country's commitment to pursuing an ambitious climate policy. By focusing on party politics, which is subject to periodic elections that may change the partisan composition of government, an explanation for variable ambition by member state governments will become apparent.

Climate change policy and EU member states

Understanding the development of the EU's climate mitigation policy has involved an analysis of inter-governmental negotiations over Commission proposals. In this context, member states' own positions may be influenced by the EU climate policy finally achieved. Having negotiated targets alters national policy, and as far as one can say that national climate policy becomes 'Europeanized', that is, an adaptation over time of national policy and policy instruments to meet EU-set goals, this has been the case for most member states.

Of course, this top-down Europeanization process is also matched by bottom-up Europeanization in which select member states are successful in uploading their preferences on various aspects of climate and energy policy to influence EU-level negotiations as well as the pre-Council phase of policy development (see Jörgens and Solorio 2017 for a discussion of Europeanization and energy policy). Nevertheless, mapping the development of member states' climate and energy policies has brought national government decision-making into perspective (see *inter alia* Compston and Baily 2008; Jordan et al. 2010; Skjærseth et al. 2016). It highlighted the variable efforts not just in terms of ambitious emission reductions but also the degree of actual national climate policy development in the first place. This appreciation of national policy positions on climate change has been an important development, as much of the literature regarding climate change politics had been focused on international bargaining and negotiations, e.g. Kyoto, various COPs and other multi-lateral gatherings such as the G20 (see Vogler 2015). But as Rootes et al. (2012) warned, 'there is a danger that the focus on the international may consign to a "black box" the politics that exists at levels below the international' (p. 678). Very little effort has gone into a comparative explanation for this variability, though Harrison and Sundstrom (2010) advanced a useful framework incorporating key factors such as different electoral systems (PR vs. majoritarian), diffusion of authority (federal or unitary) and the effect of policy lock-in by the EU. Variables such as leadership and compliance costs also feature.

As the EU's climate and energy policy developed since 2000, it is the tri-partite policy mix of carbon emission reduction, increasing the share of renewable energies, and energy efficiency that has provided a framework for national policies to adapt their own policy goals. Even more ambitious initiatives such as a European Energy Union relate to this three-part policy agenda. Consequently, analyses of national climate and energy policy have been framed by this EU agenda, providing a rough yardstick to denote climate policy 'leaders and laggards'.

Climate policy leadership

With the subsequent focus on national climate policy development, at least among EU member states, the estimation of the EU as a climate leader in international negotiations was attached to individual member states, and the terms leader and laggard began to be employed as short-hand label.[1] Informally, the use of the term leader as applied to individual member states consisted of an evaluation of a range of policies regarding climate mitigation and support for renewable energy. Indices such as the *Climate Change Performance Index* and other studies presented rankings of countries based on a mix of factors but whatever the particular rank, leaders usually consisted of countries such as Denmark, the United Kingdom and Sweden. Germany, while not in the top of rankings due to the degree of its carbon emissions, has often been categorized a leader for exercising leadership on an international level (experiences from which Chancellor Merkel derived the label of 'climate chancellor'). The concept of leadership itself has been elaborated further to distinguish between leaders and pioneers, or types and styles in an attempt to refine the meaning of leadership (Liefferink and Wurzel 2017). Nevertheless, at its most blunt, this chapter argues that although analyses of national government climate policy are indispensable, as well as defining the concept of climate leadership, further investigation of changes in leadership ranking is necessary if we are to understand the key factors that motivate governments to attain the goals they committed to in the Paris Agreement. Although Harrison and Sundstrom, as noted earlier, have usefully turned the analytical focus onto motivations to explain government policy decisions, this chapter opens up the 'black box' of national government to an even greater degree of exposure by first understanding that, at least among EU member states, national government is *party government*. Therefore, motivation of political parties, with regard to their policy position and preferences vis-à-vis climate change, is a crucial factor explaining a particular government's climate policy position and ambition.

Change of government may be the most obvious political variable explaining policy change. The most spectacular example of such change is the election of Donald Trump in the United States, demonstrating how much climate policy can change from one government to the next. Among EU member states, one might expect that the presence of a Green party with sufficient electoral strength would have a relatively strong influence, whether the party was in power or not. Party systems of a normally majoritarian construct, i.e. single party government as in the case of the United Kingdom, have also been included in the leadership category though it does not have a significant Green party. This chapter is not specifically concerned with *how* a particular country achieved leadership distinction, but rather the explanation for variation in its position or ranking over time. As mentioned earlier, implementation costs which may rise over time as the so-called low hanging fruit measures are achieved may be a factor in slowing the pace of policy development in terms of carbon emission targets or the installation of grid infrastructure to meet renewable energy goals, but the question here is how governments meet such challenges, and the party complexion of government may be a critical factor in maintaining the political will to maintain 'leadership'.

Climate policy and political parties

When we turn to political parties and policy, there is an enormous literature; however, when we narrow the focus to parties in western democracies and climate change, the literature is scant. Again, this may reflect the fact that the focus of climate policy development analysis has been concerned with the international level, yet even in descriptive accounts of national climate policy, the role of party policy on climate change is not a central concern. This section briefly presents some core issues regarding parties and climate policy, ending with a discussion of two climate country leaders, Denmark and Germany. The central line of argument is that in order to understand the complex ways in which climate policy is developed in leading countries, the contingent nature of periodic elections may have an impact on the short-to-medium term climate policy position of leading countries, that is, maintenance of leadership is not preordained.

Party competition

Climate policy, while acknowledged by voters as a significant issue in EU member states, is not a defining issue at election time, even in climate leading countries (Steentjes et al. 2017). Instead, traditional issues, such as health, jobs, education and more recently immigration, usually occupy the top five issues prioritized by voters. This being said, parties have developed positions on climate change, and within the European Union, there are a couple of notable points. First, since climate and energy policies are cross-cutting over a range of associated policies such as transport, agriculture and housing, developing a comprehensive policy is difficult as it involves many vested interests, both within and outside of parties. Therefore, for example, a centre-left party may find that promoting policies which may in the first instance increase costs to energy-intensive industry could encounter resistance from trade union allies who fear that an unintended consequence will be losing jobs. In this scenario, the trade union resistance is joined by the industry itself which may object to costs and the regulations that accompany it, therefore finding support from centre-right pro-business parties. A second point, emerging over the past 20 years or so since the higher profile of international climate negotiations (Kyoto, COP's in Copenhagen in 2009 and Paris 2015), is that climate change policy appears to fit into the domestic left vs. right pattern of party competition. Unlike the United States, where the Democrats are pro-climate policy and the Republicans all but 'anti-climate', within the EU parties of the left and right do not oppose, in principle, the fight against climate change. As was mentioned earlier, it is indeed the case that climate change is not a top priority for voters during elections – and therefore is not prominent in party manifestos or programmes – but parties have nevertheless produced positions. There are two key characteristics among parties of the left and right, including EU countries considered climate leaders. The first is that a party's ideological core identity influences particular climate policy positions; and second, in many EU political systems in which coalition government is the norm, certain constellations of party government are more conducive to climate leadership than others.

Regarding the first point, a party's position on climate-related issues can be ascertained from the manifesto employed at election time. Whether promises made during an election translate in government policy depends on (1) the nature of coalition partners' policy positions, (2) the nature of the inherited policy and possibly (3) reaction to proposed government policy, both public opinion as well as interest groups. Having stated that a party's climate position can be read in its manifesto is, however, too simple. First, most centre-left and centre-right parties in Europe are committed to fighting climate change but platitudes are one thing and detailed policy proposals another. In an analysis of the party manifestos of the two main parties of government[2] in six EU member states – France, Germany, Denmark, the United Kingdom, Italy and Ireland – it was clear that mainstream parties manifesto pledges contain both pro- and anti-climate positions (Carter et al. 2017). For example, a centre-left party may commit to boosting the percentage of renewable energy in the nation's energy supply (pro-climate) but also to build new roads (generally accepted as anti-climate). The methodology of the analysis subtracted the negative climate statements from positive ones to reach an overall score, and in so doing confirmed a hypothesis regarding left vs. right on environmental issues more generally, that centre-left parties have a stronger pro-climate policy positions. The fact that all mainstream parties have inconsistencies regarding effective climate and energy policies suggests that a more holistic approach by government has yet to emerge.

Second, except in rare cases, most party governments in Europe are composed of coalitions of two or more parties. Although the formation of a coalition government necessitates negotiations among all the potential parties in order to agree a policy programme for the duration of the government, this will mean either compromise or even some policy areas left unfulfilled. As for climate policy, the important exception to this usual scenario is the presence of an electorally significant Green party as one of the potential coalition partners. With 'blackmail potential' if no other party is available, it can insist on stronger commitments to a climate and energy transition than mainstream parties on their own. Seen from this perspective, the possibility of Green parties having such an impact on EU climate policy is small, as (1) they are electorally significant in only a handful of EU states and (2) historically they have been in coalition at a national level with only centre-left parties. They have, when in government, had a noteworthy impact, especially if we think of the so-called Red-Green governments in Germany from 1998 to 2005 and *les Verts* in France during the Hollande presidency 2012–2017. These are only two states, but importantly they are leading EU member states, and a shared commitment by these two key countries is at least favourable to EU climate and energy policy development.

Political parties form government and are responsible for implementing the policies they promised during election campaigns. Apart from Green parties, most mainstream centre-left and centre-right parties do not prioritize climate policy during campaigns although their election manifesto may contain specific references. Despite promises, the reality for most EU countries is coalition government, so the party positions are not necessarily translated easily into government policy; nevertheless, some governments have implemented climate and energy policies over the past 10 or more years that have gained them the label of 'leader'. Political parties are therefore implicated in this area of policy progress. But just as some parties in government have been able to pursue relatively strong policies while in government (qualifying as leaders), it is not impossible to conceive 'leadership' as a contested ranking or position; as mentioned earlier, party competition exists between centre-left and centre-right parties. So, are EU countries ranked as climate leaders susceptible to slowing or even reversing their leadership position? The brief case studies below outline just this type of occurrence, suggesting that leadership is not permanent and that to understand the variable nature of leadership, or ambition, understanding party policy is crucial.

Case studies in leadership and party politics: Denmark and Germany

Denmark and Germany have been labelled climate leaders since at least the late 1990s. They earned this title due to concerted efforts to reduce carbon emissions and push significantly renewable energy sources, mostly wind and solar. For Denmark, although it had begun to wean itself off of oil since the oil crises of the 1970s, it was during the 1990s that a new initiative to address climate change by focusing on renewable energy and energy efficiency took off (Andersen and Nielsen 2016; Dyrhauge 2017). Many commentators point to the critical role played by the Social Democratic Party's leader during the 1990s, Sven Auken, providing the personal leadership on climate change policy. Although not occupying the post of Prime Minister himself, his hi-profile in Danish politics, coupled with his personal commitment to climate change issues, provided the political capital for the centre-left government during the 1990s to lay the basis of Danish climate leadership. In the case of Germany, the new coalition government formed by the Social Democratic Party (SPD) and the Greens (die Grünen) in 1998, re-elected in 2002, laid the policy basis for Germany's energy transition, or *energiewende*, a legislative agenda updated by subsequent governments (Michaelowa 2008). In this case, the Greens as part of a government coalition were important in providing support for the environmentalist wing of the SPD, for as mentioned earlier, a mainstream party has multiple constituencies or interests, not all of which may support certain policies. The *energiewende* and Germany's role on the international stage in support of climate policy has represented the basis of German leadership status.

Having gained recognition as climate leaders by the early 2000s, the record shows that within the context of changing partisan government, climate policy has been influenced by the party leading a government coalition. Taking into account the finding that centre-left parties' manifestos contain more pro-climate policy positions than centre-right parties, one could expect that this should be reflected in actual government policy. Bearing in mind also that most European party governments are coalitions of two or more parties, government policy may represent a compromise on certain features of climate and energy policy, for example the source of funding for renewables, the pace of reducing carbon emissions, bringing related policy areas such as transport or heating into the wider climate and energy efforts. The focus here on climate leaders is to test this hypothesis with countries whose governments have produced the policies resulting in leadership.

Germany

Germany's international leadership is based on its support for strong international climate and energy policies, its own concerted effort at an energy transition, and finally, its position as a 'pioneer' because of its model *energiewende* (Liefferink and Wurzel 2017). Yet, although climate policy and the *energiewende* are themselves not fundamentally challenged (*á la* Trump/US Republican Party), the record since the Red-Green government coalition ended in 2005 has not been one of the linear developments, that is, progress has been variable. Climate and energy policy output from subsequent coalition governments – grand coalition of SPD and CDU (2005–2009; 2013–2017) and centre-right of FDP-CDU (2009–2013) – has struggled to advance the *energiewende* in important aspects, not the least of which is a broad realization that Germany will not meet its own 2020 climate policy target of carbon reductions. Since a new legislative package to continue the energy transition was agreed in 2015, the Green Party has lamented the modesty of government reforms, in particular the role that coal continues to play in German electricity production, a prime cause of the continuing levels of carbon emission. The ideological core of government parties helps to explain the variability. The 2009–2013 CDU-FDP coalition consisted of two pro-business parties, and despite Angela Merkel's personal commitment to climate policy, the parameters of policy were set by party compromise over policy. With the FDP having given a pronounced emphasis to its belief that reform must be market-led, it resisted direct government regulation of energy policy, especially carbon taxes and mandating the phasing out of coal. Together with like-minded CDU politicians, this coalition government carried through with previous Red-Green mandated policies but reduced its own domestic ambition with regard to new initiatives. The grand coalitions, especially from 2013 to 2017, also witnessed less progress than environmental groups (and the Green Party) expected. Here, the issue was not a coalition committed to allowing industry and business to voluntarily develop strategies towards low-carbon processes, but a coalition that itself contained a confrontation between relatively more ambitious policy (SPD) and the more cautious business friendly CDU. Although Merkel remained Chancellor and was now at more liberty to pursue a more muscular climate and energy policy, she still had elements in her party and organized business interests to contend with. On the SPD side, although it was clear that the issue of a coal phase-out was necessary, a vociferous campaign by the coal-mining union, the IGBCE, meant that the SPD energy minister felt that too direct a challenge would undermine an already weak SPD electoral base. Consequently, by 2017, a date for a coal phase-out has not been set, Germany will miss its own target for 2020 carbon reductions, and discussion is only now turning towards transport and heating as sectors whose emissions require attention to meet further reductions. Again, during this last grand coalition, Germany managed to present itself as a climate champion at international settings such as COP21 and G20, but the difference between its rhetoric and domestic accomplishments has continued to grow.

Denmark

Wind energy was promoted as an alternative for practical reasons beginning in the 1970s, not explicitly as a renewable energy response to climate change. As mentioned earlier, Danish Social Democratic-led coalitions during the 1990s initiated both the ambitious domestic programmes for renewables and energy efficiency while at the same time promoted climate policy at the international level. In 2001, a centre-right coalition government took office, led by the liberal party Venstre, which itself had adopted a low-tax, low regulation and pro-market economic policy. It set about translating this philosophy into practice within climate policy by cancelling two wind farm generators that the previous government had planned. As with the German FDP, the Venstre-led government promoted market-led progress on energy and carbon emission reductions. It appears that it was only its junior coalition partner, the Conservative People's Party, whose very popular and high-profile politician Connie Hedegaard became environment (then climate and energy) minister that the government began to emphasize climate policy, and change that came just in time for the Copenhagen COP in 2009. After the departure of Hedegaard to the European Commission and the Prime Minister to NATO in 2009, the centre-right government reverted back to the less ambitious policy of the early 2000s. The Social Democratic led coalition from 2011 to 2015 returned to more ambitious climate policy, ending its term in office with the adoption of a Climate Change Act, modelled after the UK's own 2008 legislation. Subsequently, the Venstre-led government coalition again returned to its policy of de-emphasizing government initiative in climate and energy policy, and by 2017, this lack of ambition had become reflected in international climate policy surveys, as for example: 'Experts are concerned about recent developments: the questioning and cancellation of existing climate and energy targets, and budget cuts for climate protection measures' (Burck et al. 2016, p. 5).

Conclusion

This short chapter has sought to highlight the impact that party ideology and policy positions as well as relations with parties' societal allies have on a country's position as a climate leader. While abrupt policy U-turns on climate and energy policy have not been experienced, as even less ambitious centre-right governments are still obliged, by and large, to implement a previous centre-left government policy, there is certainly an effect that partisanship plays on sustaining a country's position as a climate leader. How does the evidence of parties and climate change in this chapter impact EU climate and energy policy leadership? National efforts at carbon reduction can be facilitated through EU-wide action, for example reform of the Emissions Trading Scheme can help certain countries' industry efforts at cutting emissions. Self-interest may therefore be a driver in supporting greater or more ambitious EU policy, and by extension, global efforts. Yet, although the EU contains climate leaders, a periodic lack of ambition on their part may have a ratcheting effect on the EU's own policy development. The danger is then of complacency by the very member states that have enabled the EU to achieve its own international status as a climate leader.

Notes

1 The 'leaders and laggards' terminology originally arose from analyses of environmental policy rather than simply climate mitigation and energy policy.
2 Apart from Ireland with two major centre-right parties, the main parties of government in the other five countries were centre-left and centre-right.

References

Andersen, M. and Nielsen, H. (2016). "Denmark, Small state with a big voice and bigger dilemmas," in R. Wurzel, J. Connelly and D. Liefferink (eds), *Still Taking the Lead? The European Union in International Climate Change Politics*. Routledge, London, pp. 83–97.

Burck, J., Marten, F., Bals, C., Rink, E. and Heinze, I. (2015). *The Climate Change Performance Index, Results 2016*. Germanwatch, Bonn, Germany.

Carter, N., Ladrech, R., Little, C. and Tsagkroni, V. (2017). "Political parties and climate policy, a new approach to measuring parties' climate policy preferences." *Party Politics*, Online First, March 2017.

Dyrhauge, H. (2016). "Denmark: a wind-powered forerunner," in H. Jörgens and I. Solorio (eds), *A Guide to EU Renewable Energy Policy*. Edward Elgar, Cheltenham, UK, pp. 85–103.

Harrison, K. and Sundstrom, L. (2010). *Global Commons, Domestic Decisions, The Comparative Politics of Climate Change*. MIT Press, Cambridge, MA.

Jordan, A., Huitema, D., van Asselt, H., Rayner, T. and Berkhout, F. (2010). *Climate Change Policy in the European Union, Confronting the Dilemmas of Mitigation and Adaptation*. Cambridge University Press, Cambridge, UK.

Jörgens, H. and Solorio, I. (2017). "The EU and the promotion of renewable energy, an analytical framework," in H. Jörgens and I. Solorio (eds), *A Guide to EU Renewable Energy Policy*. Edward Elgar, Cheltenham, UK, pp. 3–22.

Liefferink, D. and Wurzel, R. (2017). "Environmental Leaders and Pioneers: Agents of Change?" *Journal of European Public Policy*, 24, 7, 951–968.

Michaelowa, A. (2008). "German climate policy between global leadership and muddling through," in H. Compston and I. Bailey (eds), *Turning Down the Heat, The Politics of Climate Policy in Affluent Democracies*. Palgrave, Basingstoke, UK, pp. 144–163.

Rootes, C., Zito, A. and Barry, J. (2012). "Climate Change, National Politics and Grassroots Action: An Introduction." *Environmental Politics*, 21, 5, 677–690.

Skjærseth, J., Eikeland, P., Gulbrandsen, L. and Jevnaker, T. (2016). *Linking EU Climate and Energy Policies, Decision-Making, Implementation and Reform*. Edward Elgar, Cheltenham, UK.

Steentjes, K., Pidgeon, N., Poortinga, W., Corner, A., Arnold, A., Böhm, G., et al. (2017). *European Perceptions of Climate Change: Topline Findings of a Survey Conducted in Four European Countries in 2016*. Cardiff University, Cardiff, UK.

Vogler, J. (2015) *Climate Change in World Politics*. Palgrave, Basingstoke, UK.

3 Diplomacy, democracy and impossible ideas

Emily Barritt

> To understand the permanent use of diplomacy and the necessity for continual negotiations, we must think of the states of which Europe is composed as being joined together by all kinds of necessary commerce, in such a way that they may be regarded as members of one Republic, and that no considerable change can take place in any one of them without affecting the condition, or disturbing the peace, of all the others.
>
> On the Manner of Negotiating with Princes, François de Callières (1716)

In Lewis Carroll's *Alice Through the Looking Glass*, the White Queen admits to Alice that she can believe as many as six impossible things before breakfast, a confession that Alice responds to with incredulity. The childlike habits of the White Queen served Alice with an opportunity to discover her own maturity, but to those attempting to craft an effective global response to climate change, it poses a challenge. It is a challenge to look to the seemingly impossible in addressing the most significant environmental problems of our times. This kind of daydreaming is already evident in green political thought, where scholars have developed Carrollian conceptions of democracy that incorporate feedback systems from plants and grant democratic rights to the unborn. In this volume on EU climate diplomacy, I respond to the royal gauntlet by considering how the rituals of EU climate diplomacy might be fashioned in similar ways to the way that democracy has been crafted, in responding to the difficulties of making just collective environmental decisions.

Connecting democracy and diplomacy

Although some accounts of environmental democracy can take surreal forms that stretch our creative capacities, it does not require a great deal of imagination to understand the basic link between democracy and diplomacy. As identified in the opening quote from François de Callières' 1716 treatise on European diplomacy, the consequences of a state's internal decisions are not contained by lines on a map and can therefore have an impact on the inner workings of other states. Thus, democratic decisions can have diplomatic consequences.

For example, the 2016 referendum on the United Kingdom's membership of the European Union will have significant consequences for the remaining members of the Union, even though they were not able to participate in the original decision. There is, therefore, an unavoidable practical connection between democracy and diplomacy.

The two ideas are also linked conceptually: both are forms of political discourse; both attempt to legitimize, negotiate and balance the exercise of power; both are guided by a set of rules and traditions about the best way to reach collective decisions; and both relate to morally preferable modes of political operation – diplomacy is better than war and democracy is better than authoritarianism or anarchy. They set high standards for our political arrangements by requiring us to act in ways that are not necessarily commensurate with our evolutionary or patriotic inclinations, to consider how human beings can operate justly in community.

In the case of climate change, the consequences of the connection between democracy and diplomacy are stark. Environmental problems, especially atmospheric ones, are not contained by borders. In the 1980s, for example, acid rain resulting from the operation of British coal-fired power stations caused significant damage to Scandinavian forests, leading to angry arguments between the United Kingdom, and Norway and Sweden. The diplomatic challenge posed by the permeability of borders is further exacerbated because of the uneven and unfair ways in which environmental problems manifest. Paradigmatic of this problem is the complete loss of territory that will be suffered by low-emitting Small Island Developing States if global temperatures continue to rise. Climate change is, therefore, a classic boundary-defying problem. As a result, profound injustices manifest on a global scale that can be broadly attributed to the democratic decisions made by high emitting states.

The central claim of this contribution is that EU climate diplomacy ought to be motivated towards achieving a just global response to the problem of climate change. Taking justice as the end goal therefore, there are ways in which climate diplomacy can and should be modified. Drawing inspiration from ecologically customized accounts of democracy, I suggest how climate diplomacy might be shaped, in tandem with environmental democracy, to achieve a more just global response to climate change.

With what follows, I intend to provoke ideas about what it means to be committed to justice in the domain of climate diplomacy. Some of the suggestions may seem obvious (entrenching the role of civil society in climate negotiations), while other suggestions may seem more outlandish (active listening to signals from the natural world). My hope is that by daydreaming the seemingly impossible, we can shape a form of EU climate diplomacy that is ambitious in its response to the injustices of climate change.

In the first section, I identify the specific difficulties associated with environmental decision-making which have led to the articulation of bespoke forms of environmental democracy and which therefore might shape climate diplomacy. Second, I sketch out the environmental democratic vision that exists in theory and has begun to emerge in practice. Third, I elaborate the various normative commitments of environmental democracy and suggest how they are relevant to climate diplomacy. Finally, I suggest two ways in which climate diplomacy might be adjusted to better account for the practical and moral challenges that arise from environmental decision-making on a global scale. In so doing, I lay the groundwork for a more just form of EU climate diplomacy that matches the commitment to environmental democracy evident in the European legal order.

Troublesome environmental decision-making

Before articulating a conception of diplomacy, that mirrors the specifically environmental conceptions of democracy that have emerged in green political thought, it is necessary to understand the features of environmental decision-making that cause it to be especially troublesome. These challenges manifest in both the domestic and global contexts, and are thus challenges that are faced by diplomatic as well as democratic processes. Accordingly, just as bespoke conceptions of democracy have emerged in response to these challenges, diplomatic practice should also be tailored to better address them, in so doing producing a more just form of climate diplomacy.

The first challenge relates to the unwieldly volume and diversity of interests impacted by environmental decisions. The nature of environmental problems is such that the decision to take an environmental risk or cause an environmental harm can impact those who would not normally be regarded as having political equality within the context of democratic decision-making, such as non-nationals, but also those who are not represented in diplomatic negotiations either, such as future generations, animals, plants and ecosystems. Thus, a commitment to just environmental decision-making might mean giving adequate attention to how the interests of these 'new environmental constituencies' can be taken into account in democratic and diplomatic decisions (Mason 1999, p. 31).

Second, the multiplicity of values involved in environmental decision-making are vast and difficult to prioritize. Value pluralism arises not just between those who advocate the primacy of environmental values and those who do not, i.e. those who favour economic rather than environmental values, but also between those who can broadly be described as environmentalists: 'equally [environmentally] conscientious citizens' can hold opposite views (Leopold 1949, p. 168). For example, an environmentally concerned citizen who wants to ensure the supply of renewable energy sources might conflict with one who is concerned to preserve the beauty of the landscape where a proposed windfarm might be placed. As a result, there is no litmus test for just or right environmental decisions. In the context of environmental democracy, deliberative processes are advocated as a means of addressing the incommensurability of environmental values. In the context of climate diplomacy, this might also require an emphasis on authentic forms of deliberation, as opposed to the more choreographed nature of diplomatic discussions.

Third, environmental knowledge claims are complex. Environmental knowledge is vulnerable to scientific uncertainty, can be context-specific and is often the product of both technical and value judgments (Fisher 2000, p. 151). Thus, environmental decisions are usually based on uncertain or complex knowledge claims, and their mixed fact/value nature means that it is not possible to prioritize 'expertise' in environmental decision-making (Beck 1995, p. 37). This is particularly challenging for diplomacy as it is ostensibly decision-making by experts. And although diplomats are experts in diplomatic process rather than technical experts, democracy is premised on a scepticism of expert decision-making, whatever the guise, because even expertise cannot produce perfect decisions, particularly where decisions involve values.

Finally, environmental risks are also especially complex and uncertain because they are based on estimates of likelihood, severity and detectability of environmental harms manifesting. When it comes to assessing environmental risks 'there are no firm probabilities', human behaviour is unforeseeable, and sometimes unthinkable, and this makes the judgment of likelihood difficult (Fisher 2007, p. 7). Further, the assessment of environmental risks is often entangled with value judgements and cultural sensitivities that also complicate how environmental decisions are made. As with environmental knowledge, experts are, therefore, of little assistance because there can be no value-free assessment of environmental risks.

It is in response to these challenges that democracy has taken on a specifically environmental complexion, for example, some accounts of environmental democracy advocate an extension of rights protection to non-humans and future generations, in order to ensure that their interests are adequately safeguarded in the context of environmental decision-making. Others promote discursive democratic processes as a way of fostering legitimacy in the context of value-laden environmental decision-making. The next section elaborates how democracy has been modified so as to respond to these challenges. In so doing, it paves the way to suggesting how diplomacy might be similarly wrought.

Sketching environmental democracy

Democracy, as with the other foundational principles of our political arrangements – justice, the rule of law, legitimacy – is impossible to contain in simple terms. It is 'an essentially contested concept' (Gallie 1955), 'an impenetrable thicket of ideas' (Dahl 2000, p. 37) and as a result, entire libraries have been written on the meaning of democracy. These manifold accounts of democracy result from different methodological and ideological choices; are motivated by different ambitions for the scope of democracy and manifest in different structural arrangements with different normative ambitions. To avoid getting caught in this thicket of ideas, I offer here an Impressionist's sketch of environmental democracy rather than a Photorealist's detailed representation.

In sketching this picture, I identify seven clear lines of democracy: the fallibility of human knowledge, political equality, autonomy, popular control, legitimacy, rights and justice. These ideas help form most, if not all, accounts of democracy and thus give an impression of the contours and shades of the concept. For example, *autonomy* corresponds to the belief that individuals should be regarded as the best judge of their own interests and that their ability to make choices about their interests should be protected within the context of collective decision-making. Whereas *popular control* recognizes that it is the 'rightness of the many, and not the few' that counts as the sole justifiable form of political system (Saward 1996, p. 81).

Democracy is, therefore, collective decision-making that counterbalances the imperfectness of human knowledge, with the political equality, autonomy and rights of individuals in a way that leads to the most just and legitimate form of popular control. *Environmental* democracy corresponds to environmental decision-making that seeks the normative authority that is bestowed by the democratic approach, i.e. collective environmental decision-making that respects the autonomy and political equality and rights of those affected by environmental decisions, in a way that recognizes the imperfectness of human knowledge and that seeks to ensure the justice and legitimacy of those decisions. However, the features of environmental decision-making identified above can make a straightforward application of democratic practice and principles challenging. Thus, distinctively environmental conceptions of democracy have emerged in response to these challenges.

To illustrate this point, consider how political equality and autonomy might be shaped by these features: environmental decisions affect a wide range of interests including those of future generations, therefore a system of environmental democracy might protect the political equality and autonomy of future generations by granting them democratic rights. Some proponents of environmental democracy go further to suggest that the democratic community should include the natural world and incorporate feedback from natural systems. That is, a commitment to environmental democracy is a commitment to 'listen to the signals emanating from the natural world with the same sort of respect we would accord communication emanating from human subjects' (Dryzek 2002, p. 149). In this way, the concept of environmental democracy can lead to some particularly outlandish suggestions for our staid political arrangements.

More modest forms of environmental democracy are already evident in international environmental law and governance. For example, the Aarhus Convention, described by then UN Secretary General Kofi Annan as an 'ambitious venture in "environmental democracy,"' commits its Contracting Parties to facilitate public participation in environmental decision-making and to guarantee access to environmental informational and justice in environmental matters. It is a concrete and meticulous expression of Principle 10 of the Rio Declaration, which introduced a basic idea of environmental democracy into international law. The focus of the Aarhus Convention on participation, accountability, transparency and the protection of basic democratic rights exemplifies a 'new era' of environmental democracy that has influenced both internal decision-making and international negotiations (Jerzy and Stec 2001).

It is, therefore, possible to think of environmental democracy in a number of different ways, as a trio of complementary procedural rights (access to information, public participation and access to justice), as a Habermasian deliberative process (Warren 1996, pp. 241–244; Eckersley 1996, p. 217) or as an eccentric practice of listening to plants (Dryzeck 2002, p. 149). While these different manifestations of environmental democracy provide inspiration for how climate diplomacy might operate in a more environmentally just manner, it is the normative core of environmental democracy that is crucial for understanding how more just climate diplomacy might be achieved. In the next section, I set out the key normative imperatives of environmental democracy and consider how they might influence EU climate diplomacy.

The imperatives of environmental democracy

Environmental democracy is motivated by the central normative concern that collective environmental decision-making should be alert to wide variety of interests and values that are affected and engaged in relation to the natural environment. It emerged as a response to the injustices that result from environmental problems, such as climate change, and as an antidote to the eco-authoritarian sentiments that permeated the environmental movement of the 20th century (Bramwell 1989). This commitment to reach just and inclusive environmental decisions can equally be applied to the context of climate diplomacy. Global decisions made (or perhaps more critically, not made) in relation to the climate will have consequences that are felt across geographical, temporal and species boundaries, therefore diplomatic efforts should be motivated, at least in part, towards addressing the injustices caused by climate change. This leads to several unifying concerns for environmental democracy and climate diplomacy that have the potential to drive a progressive form of climate diplomacy.

The first unifying concern is that environmental democracy and thus climate diplomacy ought to aim towards 'environmental goals' (Steel 2001, p. 419). This is challenging for both democracy and diplomacy. Democracy, by its very nature, does not work towards predefined goals because it is predicated on a commitment to open-ended decision-making. For diplomacy, this is challenging, because of the difficulty of reaching consensus on anything between states, and as highlighted earlier, the plurality of values involved in environmental decision-making are vast and often culturally dependent.

However, to agree to be broadly motivated towards environmental ends does not remove the open-ended quality of decision-making. Rather, the focus is about resolving the complex problems of environmental protection in as just a way as possible, considering the various interest and values associated with environmental decision-making. There already exist basic moral limitations on the consequences of democratic decisions, for example, even a democratic decision could not authorize the arbitrary imprisonment of a minority group, because certain fundamental rights and interests are put out of reach of the democratic process (Harrison 1993, p. 230). This teleological commitment shapes some of the boundary markers of democracy – political equality, autonomy, rights and justice – to give them an ecological complexion. This shaping of the boundary concepts is also relevant to diplomatic relations. For example, infusing the concept of justice with environmental considerations would require those operating within the field of EU climate diplomacy to consider the intergenerational or inter-species cost of their strategies.

The second imperative is that there needs to be a more expansive conception of the common good. The existing democratic structures have a tendency to equate the common good as a simple mathematical exercise of adding together individual goods. However, this aggregative form of collective decision-making ignores the value of certain communal goods. Environmental democracy requires a less atomistic approach to democracy that places individuals in their environmental and communal context so that these goods are part of the evaluation of the common good. A richer understanding of the collective good also requires an expansion of the democratic community, so that non-nationals, future generations and non-humans are regarded as deserving consideration of their interests in the context of collective decision-making (Doherty and de Geus 1996, p. 7).

The need to expand the democratic community to include the new environmental constituencies ought to apply in the context of diplomatic actors as well. That is, states need to think beyond their own citizenry, about the citizens of other states, future citizens and the natural world, when formulating their negotiating positions. This already happens to a certain extent within the context of EU climate diplomacy by virtue of the fact the EU is a collection of states with separate democratic identities. Nevertheless, there should still be an obligation on EU negotiators to think about the interests of those who are not privileged with EU citizenship and to consider how moral priority can be given to non-humans and future humans.

Third, long-term generalizable interests need to be given moral priority or privilege in environmental decision-making, both in a democratic and a diplomatic context. Again, this is not to dictate specific outcomes but it is to ensure that the consequences of environmental risks, which do not always materialize at the time a decision is made, are properly considered. The prioritization of long-term generalizable interests provides a necessary guide to decision-making to ensure that the interests of future generations are given more serious consideration in democratic procedures than they are in traditional representative democracies. This priority is already apparent in climate diplomacy through ideas such as the precautionary principle and the concept of sustainability, which features significantly in the preamble and main text of Paris Agreement (United Nations 2015),[1] but a commitment to just climate diplomacy requires explicit recognition of future interests, particularly as concepts such as the precautionary principle and sustainability are notoriously vague.

Fourth, because of the far reaching and even unpredictable consequences of environmental decisions, environmental democracy and climate diplomacy must strive towards improved legitimacy. The consequences of environmental decisions are often felt most keenly by those who have little or no opportunity to influence the decision, for example, indigenous communities, future generations and members of the natural world. Therefore, in order to achieve some standard of democratic legitimacy in relation to environmental decision-making, there needs to be another source of that legitimacy. For example, a strengthened position for proxies to represent the interests of democratically excluded constituencies or higher level of scrutiny of environmental decisions.

Further, there is a need to be more legitimate because of the significant role that values play in environmental decision-making, particularly in relation to environmental knowledge and environmental risk assessment which are both value-laden concepts and are therefore highly culturally sensitive. As there are few objective standards for judging which values are to be preferred in respect of environmental decisions, the adoption of certain values over others requires considerable legitimation. This is a particular challenge for EU climate diplomacy, where the democratic connection between citizens and Union negotiators is a number of steps removed. As a result, there is a pervasive, if inaccurate, perception that EU political actors, including negotiators, lack democratic legitimacy.

These four normative concerns underpin many different forms of environmental democracy that aim to produce more just environmental decisions. For example, the Aarhus Convention's focus on access to environmental information stems from a commitment to accountability and transparency, which are critical aspects of legitimacy, or John Dryzek's democratic vision that promotes alternative spheres of democratization leads to a more holistic conception of the common good (Dryzek 2000, pp. 156–157). In the final section, I suggest two ways in which these imperatives might promote a more ambitious form of climate diplomacy that is progressive in its commitment to justice.

Ambitious climate diplomacy

The influence of the international commitment to environmental democracy is already apparent in recent diplomatic responses to climate change: the decision to adopt the Paris Agreement recognized that 'ambitious climate action' required participation from non-Party stakeholders such as 'civil society, the private sector, financial institutions, cities and other subnational authorities, local communities and indigenous peoples'; the emphasis on transparency in relation to the Nationally Determined Contributions; and the existence of the High Ambition Coalition, which united states according to their ambition and not their geography, are all evidence of this. Thus, climate negotiations have already been influenced by democratic practice. However, as ever, there is more to be done if, as a European community, we want a truly just global response to climate change. In this final section, I make two suggestions for how to be truly committed to justice in climate diplomacy. First, there needs to be better recognition of those affected by climate change but who are not officially represented in either the diplomatic or democratic discourse, such as animals and future generations. Second, new diplomatic actors, cities, NGO and civil society organizations, need to have protected rights to participate in climate negotiations.

The key idea that separates environmental democracy from other progressive forms of democracy is that it envisages the bestowal of democratic privileges on constituencies not otherwise recognized in democratic processes – future generations, animals, plants, ecosystems, etc. A diplomatic process that includes official representatives for future generations and animals is a proposal that could have come straight from the imagination of Carroll himself. However, this suggestion could be achieved quite straightforwardly through my second suggestion that non-Party stakeholders such as NGOs, civil society organizations and cities should be formally involved in climate negotiations. This is because non-Party stakeholders could be invited to participate in the diplomatic process based on their commitment to the representation of interests that are not already represented in diplomatic processes.

Although the process that led to the Paris Agreement did incorporate NGOs and civil society organizations, non-Party stakeholders are still regarded as second-tier participants in the context of climate negotiations. The presence of these actors is a revocable privilege rather than an entitlement, as evidenced by the exclusion of non-Party stakeholders from certain aspects of the negotiations. This is problematic, in part because it limits transparency and accountability of decisions, essential democratic qualities, but also because it reduces adequate representation of climate victims such as future generations and the natural world. Thus, there is a need to entrench and adequately protect the role of non-Party stakeholders in climate negotiations.

A more concrete role for non-Party stakeholders is also necessary, because non-Party stakeholders are playing an increasingly active role in responding to climate change. For example, in the wake of the United States' announced withdrawal from the Paris Agreement, US city mayors rushed to reaffirm their pledges to reduce climate change. This was a new kind of democratic act within diplomatic relations. The city mayors recognized that their responsibility to the livelihoods and desires of their citizens required action, particularly in the face of federal executive inaction. Therefore, if non-Party stakeholders are playing an active role in ensuring that global targets are met, they should also be allowed to play an active role in negotiating those targets, particularly if they drive a more ambitious agenda.

The influence of the precepts of environmental democracy is already identifiable in climate change negotiations. The age of remote, privileged diplomats discussing global politics behind closed doors, is no longer an appropriate way to handle the global threat of climate change. Ambitious climate diplomacy must be open and facilitates the active representation of the great host of interests impacted by the consequences of climate change. In short, climate diplomacy needs to be more democratic.

Coda: a final daydream

Before closing, I return to the White Queen, daydreaming the impossible before breakfast. The thrust of this contribution has been that diplomacy ought to learn from democracy, implying that the traffic is only one way. Far from it. Democracy can and should learn from diplomacy. In a world where our adversarial and combative democratic processes have produced divisive results, reckless to the consequences for the future of our planet, democracy should learn the diplomatic arts of careful listening and gentle persuasion.

Note

1 Article 2(1); Article 4(1); Article 5(2); Articles 6(1), (2), (4), (8) and (9); Article 7(1) and (9); Article 8(1); and Article 10(5).

Bibliography

Beck, U. (1995). *Ecological Politics in an Age of Risk*. Polity Press, Cambridge, UK.

Bramwell, A. (1989). *Ecology in the 20th Century: A History*. Yale University Press, New Haven, CT.

Dahl, R. (2000). *On Democracy*. Yale University Press, New Haven, CT.

De Callières, F. (2000). *On the Manner of Negotiating with Princes*. Houghton Mifflin Harcourt, Boston, MA.

Doherty, B. and de Geus, M. (1996). "Introduction," in: B. Doherty, and M. de Geus (eds), *Democracy and Green Political Thought: Sustainability, Rights and Citizenship*. Routledge, Oxford, pp. 1–15.

Dryzek, J. (2000). *Deliberative Democracy and Beyond: Liberals, Critics, Contestations*. OUP, New York.

Eckersley, R. (1996). "Greening liberal democracy: the rights discourse revisited," in B. Doherty and M. de Geus (eds), *Democracy and Green Political Thought: Sustainability, Rights and Citizenship*. Routledge, Oxford, pp. 212–236.

Fisher, E. (2000). "Drowning by Numbers: Standard Setting in Risk Regulation and the Pursuit of Accountable Public Administration." *Oxford Journal of Legal Studies*, 20, 1, 109–130.

——. (2007). *Risk Regulation and Administrative Constitutionalism*. Hart Publishing, Oxford.

Gallie, W.B. (1955). "Essentially Contested Concepts." *Proceedings of the Aristotelian Society*, 56, 167–198.

Harrison, R. (1993). *Democracy: Problems of Philosophy*. Routledge, Oxford.

Jendrośka, J and Stec, S. (2001). "The Aarhus Convention: Towards a New Era in Environmental Democracy." *Environmental Liability*, 9, 3, 140–151.

Leopold, A. (1949). *Sand County Almanac: and Sketches Here and There*. OUP, New York.

Mason, M. (1999). *Environmental Democracy*. Earthscan Publications, London.

Saward, M. (1996). "Must democrats be environmentalists," in B. Doherty, and M. de Geus (eds), *Democracy and Green Political Thought: Sustainability, Rights and Citizenship*. Routledge, Oxford, pp. 79–96.

Steele, J. (2001). "Participation and Deliberation in Environmental Law: Exploring a Problem-Solving Approach." *Oxford Journal of Legal Studies*, 21, 3, 415–442.

United Nations/Framework Convention on Climate Change. (2015). *Adoption of the Paris Agreement*, 21st Conference of the Parties. United Nations, Paris, France.
Warren, M. (1996). "What Should We Expect from More Democracy?" *Political Theory*, 24, 2, 241–270.

4 The historical evolution of EU climate leadership and four scenarios for its future

Hayley Walker and Katja Biedenkopf

Introduction

The European Union (EU) has evolved into a key player in international climate governance, not just in the United Nations Framework Convention on Climate Change (UNFCCC) negotiations but also in the broader field of global climate governance, including financial support for vulnerable countries as well as bilateral and multilateral diplomatic outreach. This chapter traces the EU's historical evolution as a climate leader and ponders its future by sketching four imaginable but accentuated scenarios. The real course of action will likely be somewhere between these sketches and combine elements of different scenarios. Much depends on the United States of America (USA) and whether its change of policy under President Donald Trump will trigger a ripple effect, which could lead to a lowering of other countries' climate ambitions or to a defiant reaction by others, increasing their ambition. Yet, much also depends on the EU and its ability to weather domestic turbulence and remain a strong international climate leader. Signs of a strengthened EU–China partnership provide a glimmer of hope.

International negotiations and global governance rely on individuals and jurisdictions to take the lead and drive the process. The presidency of different negotiation rounds can provide additional impetus but ultimately, in a party-driven process, leadership should derive from the negotiating parties. Presidencies generally fulfil the role of mediators while negotiating parties can pursue ambitious positions and actively push for them, for example, by persuading others. Moreover, countries hold the key to implementation of international agreements. By adopting ambitious policies and promoting them skilfully at the international level, leaders can steer the negotiation process. With the implementation of the goals of the Paris Agreement on climate change still quite uncertain and its dependence on ever-increasing ambition levels from all countries, the demand for climate leadership has never been higher.

Actors can exert leadership in different ways: by developing innovative ideas and solutions, by implementing ambitious measures to establish an example, by convincing others to adopt ambitious practices through diplomatic skills and by using structural power to compel change (Andresen and Agrawala 2002; Young 1991; Parker and Karlsson 2010; Wurzel and Connelly 2011; Skodvin and

Andresen 2006; Malnes 1995; Underdal 1991; Liefferink and Wurzel 2017). In this chapter, we focus on *positive* leadership, which aims at increasing the level of climate policy ambition and pushing for outcomes that promote a global good (Underdal 1994). *Negative* leadership, which pursues the lowering of climate policy ambition, is not included and considered *laggardship* rather than *leadership*. Academic literature on leadership has evolved over the past two decades and recent conceptualizations differentiate four distinct types: cognitive, exemplary, entrepreneurial and structural leadership (Liefferink and Wurzel 2017). Each of the leadership types describes a different mechanism of how an actor can attract followers, thereby influencing an increase of climate policy ambition.

The four climate leadership types describe different activities. *Cognitive leadership* describes the production of innovative ideas that redefine interests and provide possible solutions. *Exemplary leadership* is the implementation of innovative ideas and ambitious measures. It can demonstrate the ambitious measures' feasibility and can boost the leader's credibility and legitimacy in international negotiations. *Entrepreneurial leadership* is exerted by convincing other actors through diplomatic skills to change their behaviour and increase their climate policy ambition so as to facilitate compromise. *Structural leadership* is derived from the position that leaders occupy in a system, which can provide them with leverage over other members of that system. The leader has power over followers and can exert pressure on them. This includes economic power but can also refer to the position of a large GHG emitter without whom an international climate agreement would not be effective (Liefferink and Wurzel 2017; Oberthür and Groen 2015).

Engaging in one single type of leadership activity could theoretically generate followers but a leader more likely employs a combination of leadership activities. The four types can be complementary and amplify each other. As outlined in this chapter, the EU strengthened its leadership performance by venturing into multiple leadership types and fortifying its initial cognitive leadership with improved exemplary and entrepreneurial leadership (Oberthür and Roche Kelly 2008; Underdal 1994, pp. 184–185). Yet, not all factors that determine the EU's successful climate leadership are under its direct control. The EU can improve its production of ambitious and innovative ideas for international negotiations (cognitive leadership), adopt ambitious domestic climate policy to lead the way (exemplary leadership), devise active persuasion and diplomatic outreach strategies (entrepreneurial leadership) and exploit its (limited) structural leadership; but the success of these activities also depends on external factors to which the EU must adapt (Oberthür and Groen 2015; Oberthür 2011).

The following section traces the historical evolution of the European Union's climate leadership, which began with cognitive leadership in the relative absence of exemplary, entrepreneurial and structural leadership. Over the course of time, the EU strengthened its credibility by adopting increasingly ambitious domestic climate policy and more recently enhanced its climate diplomacy and used its structural power as a major donor of climate finance. The chapter then proceeds to a short description of the contours of contemporary EU climate leadership before pondering possible future scenarios in a world that has been rocked by numerous shocks, most notably the economic crisis, rising populism and isolationism, and the election of Donald Trump as US President. This challenging context can provide either an opportunity for strong EU climate leadership, if it brings its domestic house in order, or an impediment to EU leadership due to domestic turbulence and international loss of momentum. The concluding section provides our interpretation of the four scenarios.

The historical evolution of EU climate leadership

Since the inception of international climate governance and the UNFCCC, European countries and increasingly the EU as a single actor have promoted ambitious positions in the various negotiation rounds (Oberthür and Roche-Kelly 2008). With regard to attracting followers, however, the EU's success has fluctuated over time as a result of both internal as well as external factors and is marked by adjustment to the changing context and learning from experience.

Cognitive leadership consistently runs through the EU's international climate role. During the negotiations to establish the UNFCCC in 1992, European countries advocated setting a legally binding treaty commitment to stabilize greenhouse gas (GHG) emissions at their 1990 levels by the year 2000. This target is indeed included in the treaty, however, in the form of an aspirational goal without legal status. The EU was forced to concede to other developed countries, particularly the USA on this demand (Bäckstrand and Elgström 2013; Oberthür 2011). This partial success was replicated five years later during the negotiations to establish the Kyoto Protocol under the UNFCCC auspices. The EU achieved its desired outcome of legally binding greenhouse gas emission reduction targets for developed countries but, to achieve this goal, it was forced to make concessions to the USA. These included the use of market-based mechanisms such as the Clean Development Mechanism, which allowed wealthy developed nations to purchase emission reduction credits from other parts of the world instead of reducing their own emissions. During the Kyoto Protocol negotiations, the EU also demonstrated early signs of exemplary leadership by accepting the largest emissions reduction target: an 8% reduction on 1990 levels by 2010 (Bäckstrand and Elgström 2013; Oberthür 2011; Van Schaik 2012).

The subsequent years marked a high point in EU leadership at the international climate negotiations. The US withdrew from the Kyoto Protocol in 2001, which created a vacuum that the EU filled by ramping up its entrepreneurial leadership. The EU is largely credited for its diplomatic efforts to ensure agreement over the implementation rules of the Kyoto Protocol, known as the Marrakesh Accords, in 2001. Once again, the EU was required to make concessions in order to retain the support of other developed countries (Bäckstrand and Elgström 2013; Oberthür 2011; Van Schaik 2012) but over the coming years the EU was praised for saving the Kyoto Protocol, which had been declared 'dead' by US National Security Advisor Condoleezza Rice (Bäckstrand and Elgström 2013). Most notably, in 2004, the EU employed structural leadership to secure the entry into force of the

Protocol by ensuring Russian ratification in exchange for EU concessions on Russia's bid to join the World Trade Organization (Bäckstrand and Elgström 2013; Oberthür 2011, Van Schaik 2012).

The Kyoto Protocol's period of legal force was due to expire in 2012, and so in the 2007 negotiations, the EU successfully pushed for the launch of a new process to establish a post-2012 agreement to be concluded by 2009 at the latest, known as the Bali Roadmap (Van Schaik 2012). The Bali Roadmap did not, however, meet all of the EU's demands. The EU had called for the creation of a global agreement that would overcome the bifurcation between developed and developing countries. In the face of considerable opposition from developing countries, the issue was put aside to be resolved at a later date (Bäckstrand and Elgström 2013; Oberthür 2011; Schunz 2015). The question remained unresolved by the 2009 deadline and the disastrous negotiations that took place in Copenhagen marked a nadir for EU climate leadership and for multilateral climate governance in general. The EU pursued cognitive leadership as in earlier negotiation rounds by advocating a legally binding agreement, the peaking of global emissions by 2020 and global emission reductions by 50% on 1990 levels by 2050.

In the advent of the 2009 Copenhagen COP, the EU had made significant progress with regard to its exemplary leadership, adding credibility to its negotiation demands. It had made considerable progress in developing and implementing comprehensive and ambitious domestic climate policy. Its decoupling of economic growth from GHG emissions and progressively more ambitious policies increased the EU's credibility on the international stage and allow it to lead by example (Oberthür and Roche-Kelly 2008; Van Schaik 2012). The EU's climate and energy package of 2008/2009 passed before the Copenhagen summit and was the most ambitious in the world at that time (Parker et al. 2017). However, this exemplary leadership was unsuccessful at Copenhagen since it did not result in attracting followers (Schunz 2015), arguably due to a lack of entrepreneurial leadership by engaging with and taking into account the other negotiating parties.

In terms of entrepreneurial leadership, the EU's ambitious position was not shared by other parties and the EU did not succeed in persuading them. The resulting Copenhagen Accord contained no emission reduction targets, had no legal force and was not even adopted by the Conference of the Parties but merely 'taken note' of as parties could not reach a consensus (Groen et al. 2012; Oberthür 2011; Van Schaik 2012). The EU found itself sidelined during the last crucial hours of the conference and did not even have a seat at the table when the final text was negotiated behind closed doors between BASIC countries (Brazil, South Africa, India and China) and the USA that had newly reengaged in the process under the Obama administration (Bäckstrand and Elgström 2013; Herrero and Knaepen 2014; Parker et al. 2017; Schunz 2015). The EU failed to adapt to the new world order in which countries like China and India had risen. It attempted to employ structural leadership through offering financial incentives to countries in exchange for accepting its vision for post-Kyoto climate governance and it attempted to employ exemplary leadership through its ambitious pledges. However, its diminished position in the global climate system and its decreasing share of global GHG emissions undermined these modes of leadership.

The EU's capacity to speak with one voice at multilateral negotiations and conduct itself as a unified actor had been a challenge in the earlier climate negotiations and was identified as one of the contributing factors to the EU's failure as an entrepreneurial leader at the Copenhagen COP. The demands of internal coordination meant that the EU was unable to spend sufficient time conducting climate diplomacy and other forms of outreach to negotiation parties, which would have been necessary to influence, broker and form coalitions, thus exerting entrepreneurial leadership (Oberthür 2011). In Copenhagen, divergent preferences among member states caused conflict on certain issues (Groen et al. 2012; Schunz 2015) and the usual system of cohesive coordination between environment ministers broke down when the negotiations moved to the level of heads of state (Keukeleire and Delreux 2014), who failed to coordinate both among themselves and with the Swedish EU Presidency (Van Schaik 2012).

Recognizing its failure to act as entrepreneurial leader, the EU improved its climate diplomacy strategy over the following years, acknowledging that an elaborate outreach strategy that engages with and understands other negotiation parties' positions is crucial. The EU started adapting to the new state of affairs in global climate politics and engaged in coalition building with progressive developed and developing countries (Bäckstrand and Elgström 2013; Biedenkopf and Walker 2016; Oberthür and Groen 2017; Parker et al. 2017). The EU is credited with modest goal achievement in the negotiations that took place the following year and resulted in the Cancun Agreements (Groen et al. 2012; Oberthür 2011; Schunz 2015; Van Schaik 2012) and with more significant success at the 2012 negotiations which launched the Durban Platform, a new negotiation track with the aim of adopting a global agreement applicable to all parties by 2015 (Bäckstrand and Elgström 2013; Parker et al. 2017; Schunz 2015). In exchange for this success, the EU was required to reverse its earlier refusal to sign on to a second commitment period of the Kyoto Protocol, which was on the brink of expiry. This was considered by some as the 'recovery of the EU's leadership after Copenhagen' (Bäckstrand and Elgström 2013, p. 1380) due to its critical role as entrepreneurial leader through building bridges between opposing camps. This new role for the EU was applied successfully in the run-up to and during the 2015 negotiations in Paris on a new global agreement. The EU built alliances with other progressive nations that came to be known as the *High Ambition Coalition* and which is credited with raising the level of ambition of the final outcome (Oberthür and Groen 2017). The Paris Agreement reflected most of the EU's preferences, yet in a compromise version: it is a global treaty with mitigation commitments from all parties, and although the process of submitting increasingly ambitious mitigation targets is binding, the achievement of these targets is not legally binding as the EU had wished.

The EU's climate leadership has fluctuated over the course of time and is marked by adjustment to the changing context and learning from experience.

The EU's historical climate leadership has been shaped not only by its own performance as an actor, but also by other key actors, most notably the USA. The USA first withdrew from the Kyoto Protocol and then reengaged in the UNFCCC process. The effect of this on EU climate leadership was profound in both cases. In the first case, US withdrawal not only left a leadership vacuum for the EU to fill, but also left the EU with by far the greatest share of emissions under the Kyoto Protocol. It held the most structural power under the bifurcated system. This period coincided with a peak of EU climate leadership, as it was credited with saving the Kyoto Protocol and ensuring its entry into force. This served to mask the effects of the EU's falling share of global emissions and the rising influence of emerging economies and developing countries in a global multipolar system of climate politics. When the USA reentered the process, the opposite effect was felt and the EU's declining relevance was amplified in Copenhagen (Oberthür 2011). In the run-up to Paris, the EU and the USA engaged in complementary climate leadership activities, mutually supporting each other's positions to a large extent (Biedenkopf and Walker 2016).

The contours of contemporary EU climate leadership

The EU has evolved and reshaped its leadership in a continuous learning and adaptation process in the international climate negotiations. Internal institutional developments and politics have also contributed to shaping the contours of EU climate leadership. The system of internal coordination, the EU's approach to climate diplomacy and domestic EU climate policy have matured but face challenges in the wake of Brexit and with the lowering of climate ambition in some EU member states. This occurs in a turbulent external context marked by the USA's recent disengagement from international climate governance and its possible consequences.

The EU has streamlined its internal coordination for international representation at the climate negotiations, which is marked by formal and informal features. The European Commission, more precisely its Directorate-General for Climate Action, plays a key role and represents the EU together with the Council whose presidency rotates every six months. To ensure continuity within the Council, a system of lead negotiators and issue leaders was introduced by the 2004 Irish presidency (Oberthür 2011; Van Schaik 2012). The Working Party on International Environmental Issues (WPIEI), expert groups and the so-called *EU Team*, which comprises different lead negotiators and experts from both the Commission and member states are key actors of EU climate diplomacy. The expert groups prepare technical aspects of the EU's position for international negotiation rounds. The WPIEI is an official Council working group and chaired by the EU presidency. It supervises the expert groups and determines the EU's position that builds the basis for the EU Team's work. The EU Team includes the lead negotiators, track coordinators and issue leaders and negotiates on behalf of the EU, except when the negotiations shift to the ministerial level. At this stage, the Commissioner for Climate Action and the responsible minister from the country holding the EU presidency represent the EU (Delreux 2018). This system enables the EU to negotiate more efficiently than in earlier times. The lower reliance on the EU presidency that rotates every six months and the close cooperation among Commission and Council actors have improved the EU's ability to produce positions ahead of negotiation rounds, speak with one voice and act in a coherent manner. The EU has evolved into an actor that is able to develop ideas and positions, which are the basis for its long-term strength as cognitive leader.

The EU has also taken steps to elaborate a coherent climate diplomacy approach in support of the UNFCCC negotiations. The European External Action Service (EEAS) created by the Lisbon Treaty has developed various climate diplomacy strategies jointly with the Council and the Commission (Council of the European Union 2011, 2013, 2016; European Commission and EEAS 2015; EEAS and European Commission 2011, 2013). In 2012, the EEAS took over the informal *Green Diplomacy Network* that comprised representatives of member state foreign ministries and Commission officials and had originally been designed to mainstream environmental issues into EU member states' diplomacy (Torney and Davis 2018). In spite of capacity restraints and limited resources for climate diplomacy both at the EEAS and in the EU delegations around the world (Herrero and Knaepen 2014), the EU was able to conduct comprehensive and effective outreach in the run-up to the Paris summit that contributed to mobilizing the High Ambition Coalition. Through activities such as the *Climate Diplomacy Day* and *démarches*, a large number of EU Delegations together with member state embassies actively reached out to their host country in support of the EU's negotiation position, fostering support for the international climate negotiations. Coordination practices at third-country level among EU Delegations and member state embassies have been established, yet largely function as information-exchange fora rather than groups that develop joint strategies. In light of these changes, of all the types of leadership, the EU has arguably made the largest leap forward in its entrepreneurial leadership.

In addition, outreach to developing countries is marked by the mainstreaming of climate action in EU development aid, translating into structural leadership. The EU committed to dedicate 20% of its 2014–2020 expenditures to climate-related objectives. Applied to development aid, this signifies an estimated 14 billion Euros of climate-related development cooperation financing (Adelle et al. 2018).

Internal EU climate policy has matured over the past two decades with the ambitious goals for 2030 to reduce GHG emissions by at least 40% compared to 1990 level, source at least 27% of the EU's energy from renewable sources and improve energy efficiency by at least 27%. In contrast to the 2020 EU climate targets, the 40% emission reduction goal excludes offsets and is confined to domestic efforts. A number of legislative proposals to set the EU on track to achieve its goals are currently in the political process. Yet, the politics of the EU's climate policy-making process have proven challenging. The global financial and Eurozone crises of 2008 shook European economies and led to austerity measures and reluctance to adopt ambitious climate policies. The United Kingdom's exit of the EU will remove a strong and influential supporter of ambitious EU climate policy, tilting the internal EU power balance towards less ambitious member states. Poland has emerged as the leader of a coalition of member states that strive to lower the EU's climate ambition and roll back climate policy (Burns and Tobin 2018). These developments pose increasing difficulties for the EU's signature exemplary leadership.

Additionally, the external constellation of international climate politics has undergone significant changes during the lifetime of the UNFCCC negotiations. The relative position of the EU within this system has a profound effect both on its leadership potential and types available to it. Structural power in climate politics partially derives from one's share of global GHG emissions (Bäckstrand and Elgström 2013). In 1990, the EU was responsible for about 24% of global emissions; this figure has since fallen to under 10%. In contrast, China's share of global GHG emissions has risen over the same period to about 30% and it is joined by other emerging economies that are set to overtake the EU's share of global emissions. These developments have fundamentally altered the power balance in the international negotiations.

This trend has led to another fundamental shift in global climate politics. In the early days of the UNFCCC and the Kyoto Protocol, a clear bifurcation existed between developed countries, who held responsibility for the burden of climate change, and developing countries who were not subject to any formal obligations. As developing countries continued to increase their share of global GHG emissions, it became clear that action limited to developed countries would not be sufficient and a new paradigm began to evolve in which developing countries would be expected to contribute, and as a result would have a greater say over negotiation outcomes (Bäckstrand and Elgström 2013; Oberthür 2011; Schunz 2015). This context increases the need for entrepreneurial leadership and reduces the EU's capacity to act as structural leader.

These contours of contemporary EU leadership do not point coherently into the same direction. The tremendous institutional and climate policy progress that the EU has achieved points towards continued strong leadership, while the growing internal contestation of ambitious climate policy and the rapidly changing international context point towards diminishing EU leadership. Based on these contradictory signs, we develop four scenarios for future EU climate leadership, all of which are accentuated and exaggerated. They are intended to serve as thought starters for a discussion on how to shape the EU's future role in international climate governance.

Future scenarios for EU climate leadership

The four scenarios take the factors and developments identified in the previous two sections as a starting point and sketch possible future courses of action. The first two scenarios paint a picture of EU leadership loss, while the last two describe strengthened EU leadership. Scenario 1 ponders the risk of internal EU collapse and scenario 2 the risk of the lowering of ambitions by external actors. Contrary to this, scenario 3 imagines stronger EU leadership unilaterally and scenario 4 joint strong EU–China leadership.

Scenario 1: internal collapse curtails the EU's international role

Domestically, the EU has weathered many crises that threatened to undermine its internal cohesion and leadership capacity at international climate negotiations. One possible future scenario could be a collapse of internal EU ambition as a result of Brexit, economic concerns taking precedence and laggard member states gaining influence. Since 2008, Poland has emerged as the spearhead of a group of EU member states that aim to lower the ambition and scope of EU internal climate policy. Poland will preside over the crucial 2018 international climate negotiations during which the rulebook for the implementation of the Paris Agreement should be decided upon and will host the conference in the coal-mining town Katowice (Mathieson 2017a). Without the United Kingdom as influential proponent of strong climate policy, the Nordic countries, Germany and France, might not be powerful enough to maintain EU climate policy at the level that is required to achieve the 2030 goals. Germany also faces domestic challenges balancing its exit from nuclear electricity, while at the same time reducing coal, an industry with significant political influence. Germany could lower its stance on ambitious climate policy since it encounters difficulties reaching its own targets and tends to protect its polluting car industry. This would result in the loss of the EU's exemplary leadership with severe repercussions also for its entrepreneurial and cognitive leadership. Without walking the talk, the EU's credibility would be lost and with this its influence on other countries. The lowering of domestic ambition would also lead to the scaling back of climate finance and diplomacy efforts as part of the EU's external action and development cooperation. The collapse of internal EU climate policy would trigger the implosion of the EU's international climate leadership.

Scenario 2: international domino effect creates unsurmountable hurdles

EU climate leadership is threatened on the one hand by internal implosion and on the other by the explosion of external dynamics, making the context too challenging for the EU to exert leadership alone without sufficient allies. On 1 June, 2017, President Trump announced the USA's withdrawal from the Paris Agreement and has since sent mixed messages with regard to this intention. The fact that the USA sent a reduced delegation to the 2017 UNFCCC negotiation rounds in Bonn indicates that it does not intend to play a significant role going forward in climate governance (Mathiesen 2017b). This pairs with the roll back of domestic US climate policy and the hostile climate-sceptic rhetoric of the US government. US disengagement could have knock-on effects. Other countries could fear losing competitiveness and as a result decide to lower their climate ambition, normalizing this kind of behaviour. The USA's exemplary laggardship could find followers in the large group of moderately ambitious countries with strong domestic resistance from certain industry groups. Especially, if large economies such as the BASIC nations delay their commitments under the Paris Agreement and backtrack on their pledges in response to the USA's withdrawal, the EU would not be able to singlehandedly turn around international dynamics. This international political context would pose too high a hurdle for any type of EU climate leadership to exert the necessary influence to achieve the goals set by the Paris Agreement.

Scenario 3: rising to the international challenge

The USA's withdrawal from the Paris Agreement and disengagement from the international climate negotiations could equally have the opposite outcome to the negative knock-on effect outlined in scenario 2. European Commission President Jean-Claude Juncker said in his 13 September, 2017 State-of-the-Union speech: 'Set against the collapse of ambition in the United States, Europe will ensure we make our planet great again'. While the USA prepares its exit from the Paris Agreement, the rest of the world could remain committed. The EU could get its act together domestically and fill the leadership vacuum created by the USA's disengagement. Similar to the saving of the Kyoto Protocol, US disengagement could act as an external push to unite European countries and provide impetus for all EU member states to recognize and take seriously their responsibility for the future of the earth. In this scenario, climate policy could be one of the policy areas that binds all EU member states together and enables them to develop a vision for the future of Europe. The EU would need to rise to the challenge of filling the vacuum created by the USA mainly through cognitive, entrepreneurial and exemplary leadership since its structural leadership resulting from its share of global GHG emissions continues to fall, but structural leadership via climate and development aid remains part of the EU's toolbox. Forming alliances and building bridges, which have delivered success in recent years, seem crucial in combination with ambitious domestic climate policy and ambitious visions for the international level.

Scenario 4: strengthened EU–China partnership

Singlehanded climate leadership seems the most ambitious scenario. A variant of this continued EU climate leadership scenario is a strengthened EU–China partnership to jointly drive global climate ambition and fill the US-induced vacuum. Recent activity including the publication in 2016 of an ambitious 5-year plan outlining its domestic climate action indicates that China might be ready to assume the mantle of climate leader (Hart et al. 2017). At the same time as the USA announced its withdrawal, the EU and China convened a leaders' summit, where they made clear their intention to cooperate in implementing the Paris Agreement (Gaventa 2017). On 15–16 September, 2017, the EU, China and Canada convened in Montreal a meeting at ministerial level to maintain the momentum of the Paris Agreement. Thirty-four countries participated. These developments suggest the possibility of a strengthened and crucial leadership role for the EU in partnership with China and possibly some other countries. This would also be grounded in the exercising of a mix of cognitive, exemplary and entrepreneurial leadership by both the EU and China. As the largest single emitter of greenhouse gases, China would also bring significant structural power to the partnership.

Conclusions

The real future will most likely be a combination of elements from the four scenarios. With the implementation of the goals of the Paris Agreement still very uncertain and its dependence on ever-increasing levels of ambition from all countries, the demand for climate leadership has never been higher. Whether the EU can supply the required leadership will depend on a number of factors, not least the global constellation of climate politics and whether the EU is able to devote less time to introspection and more time to observing and adapting to the actions of other players.

In order to ensure the credibility and legitimacy necessary for climate leadership, the EU will need to overcome the political turbulence it is experiencing and bring its house in order. It will also need to deliver on its pledges and continue to implement deep reductions in domestic emissions. This will not be an easy task, given that the low-hanging fruits of emissions reductions have already been harvested. Further mitigation measures will require greater effort and structural changes, yet the EU's past record of learning from experience and adapting to changing circumstances gives cause for optimism.

Nonetheless, the fate of EU climate leadership will also be determined by factors outside of its control; namely its relative position in the international constellation of climate politics. The positions and (lack of) leadership assumed by the USA and China over the coming years will be of critical importance as the details of the Paris Agreement are first negotiated and then implemented. In November 2017, the first COP was held since the USA signalled their intention to withdraw from the Paris Agreement. French President Macron attempted to fill the leadership vacuum by pledging to replace the USA' financial contribution to the IPCC. At the same time, the EU was criticized for its failure to ratify the second commitment period of the Kyoto Protocol due to Poland's opposition. Germany was hamstrung by domestic negotiations to form a new federal government. Rather than displaying unity, the EU and China found themselves on opposing sides when negotiating the weighty questions concerning how the provisions of the Paris Agreement will apply to developed and developing countries, respectively. China exerted its leadership rather to push for the interests of mid-income countries than for the global climate goals. Meanwhile, a disengaged US administration that failed to provide its delegation with instructions for the COP allowed China the opportunity to assert itself more forcefully and win key victories that will hold developed countries to account for the mitigation commitments they have already made and for the levels of climate finance made available for developing countries.

While this chapter focuses on state actors, the 2017 climate negotiations in Bonn again highlighted the growing leadership ambition of other actors such as subnational states and cities. A coalition of US subnational actors, NGOs and businesses, led by California governor Jerry Brown and former New York mayor Michael Bloomberg, pledged their continued commitment to the Paris Agreement goals and expressed their leadership ambition. Such actors and developments could play into the hands of continued strong EU leadership.

References

Adelle, C., Delputte, S., De Roeck, F. and Nicholson, S. (2018). "Environmental instruments in development cooperation: promoting better development and environmental outcomes?" in: C. Adelle, K. Biedenkopf, and D. Torney (eds), *European Union External Environmental Policy. Rules, Regulation and Governance Beyond Borders*. Palgrave Macmillan, Cham, Switzerland.

Andresen, S. and Agrawala, S. (2002). "Leaders, Pushers and Laggards in the Making of the Climate Regime." *Global Environmental Change* 12, 41–51.

Bäckstrand, K. and Elgström, O. (2013). "The EU's Role in Climate Change Negotiations: From Leader to 'Leadiator'." *Journal of European Public Policy* 20, 10, 1369–1386.

Biedenkopf, K. and Walker, H. (2016). *Playing to One's Strengths: The Implicit Division of Labor in US and EU Climate Diplomacy*. AICGS Policy Report 64, Johns Hopkins University, Baltimore, MD.

Burns, C. and Tobin, P. (2018). "The limits of ambitious environmental policy in times of crisis," in: C. Adelle, K. Biedenkopf and D. Torney (eds), *European Union External Environmental Policy. Rules, Regulation and Governance Beyond Borders*. Palgrave Macmillan, Cham, Switzerland.

Council of the European Union. (2011). *Council Conclusions on EU Climate Diplomacy*. Foreign Affairs Council meeting, Brussels, Belgium, 18 July 2011.

Council of the European Union. (2013). *Council Conclusions on EU Climate Diplomacy*. Foreign Affairs Council meeting, Luxembourg, 24 June 2013.

Council of the European Union. (2016). *Council Conclusions on European Climate Diplomacy after COP21*. Foreign Affairs Council meeting, Brussels, Belgium, 15 February 2016.

Delreux, T. (2018). "Multilateral Environmental Agreements: a key instrument of global environmental governance," in: C. Adelle, K. Biedenkopf and D. Torney (eds), *European Union External Environmental Policy. Rules, Regulation and Governance Beyond Borders*. Palgrave Macmillan, Cham, Switzerland.

Elgström, O. (2007). "Outsiders' Perceptions of the European Union in International Trade Negotiations." *Journal of Common Market Studies* 45, 4, 949–967.

European Commission, and European External Action Service. (2015). *Action Plan for Climate Diplomacy*. Brussels, Belgium, January 2015.

European External Action Service, and European Commission. (2011). *Joint Reflection Paper. Towards a Renewed and Strengthened EU Climate Diplomacy*. Brussels, Belgium, 9 July 2011.

European External Action Service, and European Commission. (2013). *EU Climate Diplomacy for 2015 and Beyond. Reflection Paper*. Brussels, Belgium, 26 June 2013.

Gaventa, J. (2017). "EU–China climate statement is a manifesto for a new global order." *Climate Home* www.climatechangenews.com/2017/06/02/eu-china-statement-manifesto-new-global-order/ 22 September 2017.

Groen, L., Niemann, A. and Oberthür, S (2012). "The EU as a Global Leader? The Copenhagen and Cancun UN Climate Change Negotiations." *Journal of Contemporary European Research* 8, 2, 173–191.

Hart, C., Jiayan, Z. and Jiahui, Y. (2017). "Mapping China's Climate Policies." *China Carbon Forum.*

Herrero, A. and Knaepen, H. (2014). "Run-up to 2015: A moment of truth for EU external climate action?" *European Centre for Development Policy Management* Briefing Note 67.

Keohane, R. and Victor, D. (2011). "The Regime Complex for Climate Change." *Perspectives on Politics* 9, 1, 7–23.

Keukeleire, S. and Delreux, T. (2014). *The Foreign Policy of the European Union.* Palgrave Macmillan, Basingstoke, UK.

Kilian, B. and Elgström, O. (2010). "Still a Green Leader? The European Union's Role in International Climate Negotiations." *Cooperation and Conflict* 45, 3, 255–273.

Liefferink, D. and Wurzel, R.K.W. (2017). "Environmental Leaders and Pioneers: Agents of Change?" *Journal of European Public Policy* 24, 951–968.

Malnes, R. (1995). "'Leader' and 'Entrepreneur' in International Negotiations: A Conceptual Analysis." *European Journal of International Relations* 1, 87–112.

Mathiesen, K. (2017a). "UN Climate Conference 2018 Heads to Heartland of Polish Coal." *Climate Home* www.climatechangenews.com/2017/06/01/un-climate-conference-2018-heads-heartland-polish-coal/ 22 September 2017.

Mathiesen, K. (2017b). "US Sends 'Much Smaller' Team to Climate Talks in Bonn." *Climate Home* www.climatechangenews.com/2017/05/08/us-sends-much-smaller-team-climate-talks-bonn/ 22 September 2017.

Oberthür, S. (2011). "The European Union's Performance in the International Climate Change Regime." *Journal of European Integration* 33, 6, 667–682.

Oberthür, S. (2016). "Reflections on Global Climate Politics Post Paris: Power, Interests and Polycentricity." *The International Spectator* 51, 4, 80–94.

Oberthür, S. and Groen, L. (2017). "Explaining Goal Achievement in International Negotiations: The EU and the Paris Agreement on Climate Change." *Journal of European Public Policy*. DOI: 10.1080/13501763.2017.1291708.

Oberthür, S. and Roche Kelly, C. (2008). "EU Leadership in International Climate Policy: Achievements and Challenges." *The International Spectator* 43, 35–50.

Oberthür, S. and Groen, L. (2015). "The Effectiveness Dimension of the EU's Performance in International Institutions: Towards a More Comprehensive Assessment Framework." *Journal of Common Market Studies* 53, 1319–1335.

Parker, C.F. and Karlsson, C. (2010). "Climate Change and the European Union's Leadership Moment: An Inconvenient Truth?" *Journal of Common Market Studies* 48, 923–943.

Parker, C.F., Karlsson, C and Hjerpe, M (2015). "Climate Change Leaders and Followers: Leadership Recognition and Selection in the UNFCCC Negotiations." *International Relations* 29, 4, 434–454.

Parker, C.F., Karlsson, C. and Hjerpe, M. (2017) "Assessing the European Union's Global Climate Change Leadership: From Copenhagen to the Paris Agreement." *Journal of European Integration* 39, 2, 239–252.

Schunz, S. (2015). "The European Union's climate change diplomacy," in: J. Koops and G. Macaj (eds), *The European Union as a Diplomatic Actor*. Palgrave Macmillan, Basingstoke, UK.

Skodvin, T. and Andresen, S. (2006). "Leadership Revisited." *Global Environmental Politics* 6, 13–27.

Torney, D., and Cross Davis, M.K. (2018). "Environmental and climate diplomacy: building coalitions through persuasion," in: C. Adelle, K. Biedenkopf and D. Torney (eds), *European Union External Environmental Policy. Rules, Regulation and Governance Beyond Borders*. Palgrave Macmillan, Cham, Switzerland.

Underdal, A. (1991). "Solving collective problems: notes on three modes of leadership," in: I. Shihata (ed.), *Challenges of a Changing World: Festchrift to Willy Østeng*. Lysaker, Norway.

Underdal, A. (1994). "Leadership theory," in: I.W. Zartman (ed.), *International Multilateral Negotiation – Approaches to the Management of Complexity*. Jossey Bass, San Francisco, CA.

Van Schaik, L. (2012). "The EU's disappointing leadership on climate, or should we be less judgemental?" in: K.E. Jørgesen and K.V. Laatikainen (eds), *Routledge Handbook on the European Union and International Institutions: Performance, Policy, Power*. Routledge, Abingdon, UK.

Wurzel, R. and Connelly, J. (2011). *The European Union as a Leader in International Climate Change Politics*. Routledge, Abingdon, UK.

Young, O.R. (1991). "Political Leadership and Regime Formation: On the Development of Institutions in International Society." *International Organization* 45, 281–308.

Section 3

Law

5 The relationship between domestic and EU policy developments

The Greek case as a paradigm on climate change and clean energy

George Dellis and Eugenia Giannini

The Greek case on climate change and clean energy in a nutshell

Over the past two decades, climate issues have evolved into a major force of change in a global context. The scope of the present volume is to highlight this aspect at a domestic/EU combined level where politics, technology and networks have created a background of radical change in law. The results of such change have proved at least one certain thing: that 'long is the road and hard'. Success and failure are interchanging with such rapidity that it is hard to assess and even harder to achieve a stable legal regime.

Greece could not but follow that path in an EU regulated framework. Over the past decade, a new climate change and clean energy framework has evolved at a domestic context, following the steps of the EU and other member states. The background is familiar: 20–20–20 targets and their ambitious character shaped the policy and defined legislative choices. Greece transposed EU legislation in a rather speedy manner, at least compared to other sectors. It is also undisputed that such transposition did not trigger any negative reaction by EU institutions and for the purposes of the present analysis is presumed to be aligned with EU law. This immediate regulatory response indeed paved the way for a swift transition to clean energy and a massive explosion in the exploitation of RES. Although this might raise eyebrows, one can say that the Greek performance has even proved better in comparison with other member States. Greece is on a track to achieve the 20–20–20 targets, despite its historic close dependence on coal and its unique system characteristics with islands remaining non-interconnected with the mainland.

However, before regarding this case as a success story in the east Mediterranean we will, unfortunately, prove quite the opposite, for one must also take a close look at the loopholes of the Greek system that undermine such an optimistic impression, both at a domestic and at an EU level. Greece is gifted with considerable resources in RES, particularly in the form of wind and solar power.

Sun is undoubtedly the major advantage Greece is presented with, comparing with most EU member states. However, it is solar power that gave rise to most difficulties in the newly founded scheme for RES. Failures in clean energy politics relate primarily to the financial leg of the RES promotion strategy, especially in the solar sector.

The present analysis considers the RES status in Greece and the actual impact of regulation applied over the past decade. As will be explained, despite the extended introduction of wind and solar farms, Greece failed in achieving a smooth, coherent, economically efficient and finally sustainable transition to RES. Such a failure raises questions regarding the characteristics and the prospects of the EU's climate policy. In other words, is the financial crash of the Greek RES scheme merely another Greek-made tragedy or is it partly attributed to EU rigidness and hesitance to proceed with a holistic EU energy integration policy based on market-oriented tools?

The domestic scheme: RES incentives

The Greek Constitution recognizes the human right to the environment and establishes the principle of sustainability. According to article 24 para. 1, the protection of the environment constitutes a fundamental obligation for the State to take appropriate preventive and remedial measures. Energy is not expressly referred to in the Constitution; nevertheless, it falls under the notion of *service public* and according to the Greek Supreme Administrative Court, since it is a service of vital importance for the public, the State bears a fundamental duty to guarantee its constant provision (Council of State, opinion 57/1992). Therefore, clean energy could be qualified as a combination of two constitutional duties. However, up until recently, Greece lacked an internal RES policy, the latter being put into place only in the context of transposing the relevant EU secondary legislation.

The beginning of RES incentives: feed-in tariffs introduced

A first step in building a RES promotion scheme was taken in adopting Act 3468/2006, transposing Directive 2001/77/EC.[1] Act 3468/2006 introduced incentives for the generation of electricity from RES in a rather simple way: it granted them priority in accessing the domestic electricity market. Apart from the 'priority' rule, the Greek legislature had to tackle the problem of RES energy remuneration. The mechanisms that Member States were entitled to adopt in order to support RES penetration were two: (1) the feed-in tariffs (FiTs) system according to which the price of energy produced by RES is guaranteed and does not depend on the fluctuation of the marginal price of the energy system; and (2) the Auctions System according to which producers are licenced to produce RES energy after a bidding procedure.

Act 3468/2006 introduced the system of FiTs. According to this Act, FiTs categories are designated on several criteria, primarily that of RES technology

Table 5.1 FiTs, as initially introduced in Act 3468/2006

Energy production from	*Interconnected network €/MWh*	*Non-interconnected network €/MWh*
Wind farms	73.0	73.0
Offshore wind farms	90.0	90.0
Hydroelectric power stations < 15 MWe	73.0	84.6
Photovoltaic installations ⩽ 100 kW	450	500
Photovoltaic installations > 100 kW	400	450
Solar thermal power stations ⩽ 5 MWe	250	270
Solar thermal power stations > 5 MWe	230	250
Geothermal power, biomass	73.0	84.6
Other RES	73.0	84.6
Cogeneration of high efficiency heat and power	73.0	84.6

The tables presented are part of E. Giannini's PhD Thesis '*The penetration of RES in the energy system of Greece and the Sustainability Principle*', National Technical University of Athens (2017), based on data published by the Greek Operator of the electricity market (LAGIE, in Greek: '*ΛΑΓΗΕ*') and the IEA.

and installation capacity (within the same RES technology type). The initial FiTs, as stipulated in Act 3468/2006, are depicted in Table 5.1.

It is obvious that the Greek scheme was more favourable towards Photovoltaics (PV) technology; the average price per MW for this technology is on average six times higher than the price for Wind, Geothermal and Biomass technology. Such variation is not justified on the grounds of key costs return as illustrated below.

In the following bar chart (Figure 5.1), FiTs per technology are presented.

Act 3468/2006 has been amended several times. Regarding FiTs, Act 3851/2010 transposing Directive 2009/28/EC[2] brought about considerable changes. The major difference between Directive 2001/77 and Directive 2009/28

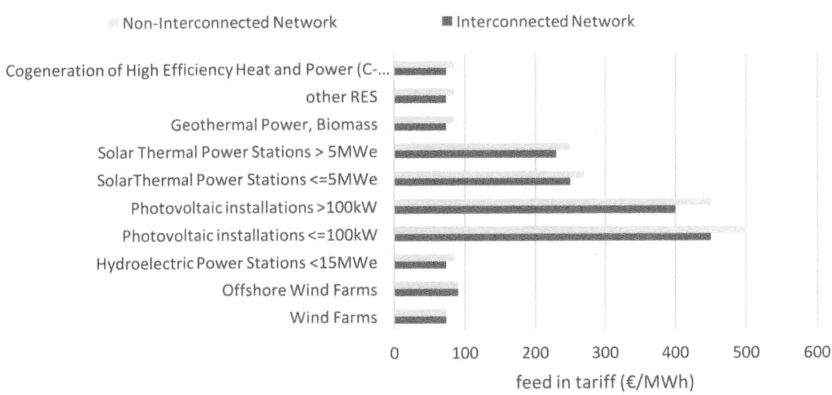

Figure 5.1 FiTs per technology compared, as initially introduced in Act 3468/2006

is that the latter introduces mandatory national targets as its major RES promotion mechanism. To meet this objective, the Greek legislature increased wind energy FiT comparatively to PV energy FiT: tariff increase for wind technology is on average 20%. On the contrary, FiTs for PV technology were put under a mechanism for a gradual decrease on a bi-annual basis. A considerable support by increase of tariffs is adopted for small capacity wind (under 50 kW) as the new tariff is three times higher than the one adopted by Act 3468/2006, in an effort to support small-scale wind projects by giving financial incentives to small-sized investments (Table 5.2; Figure 5.2).

FiTs for RES-generated power are higher than the weighted price paid to non-RES energy producers. According to Act 4001/2011, the domestic energy market

Table 5.2 FiTs per technology, as amended in Act 3851/2010

Energy production from	*Interconnected network €/MWh*	*Non-interconnected network €/MWh*
Wind farms > 50 kW	87.85	99.5
Wind farms < 50 kW	250	250
Photovoltaic installations < 100 kW	550	550
Hydroelectric power stations < MWe	87.9	87.9
Solar-thermal power stations	265	265
Solar-thermal power stations with storage capacity	285	285
Geothermal power (low T)	150	150
Geothermal Power (high T)	99.5	99.5
Biomass < 1 MW	200	200
Biomass > 1 MW < 5 M	175	175
Landfill gas (biological cleaning < 2 MW)	120	120
Landfill gas (biological cleaning > 2 MW)	99.5	99.5
Biogas (biomass < 3 MW)	220	220
Biogas (biomass > 3 MW)	200	200
Cogeneration of high efficiency heat and power (C-HE-HP)	87.9	99.5
Other RES	87.9	99.5

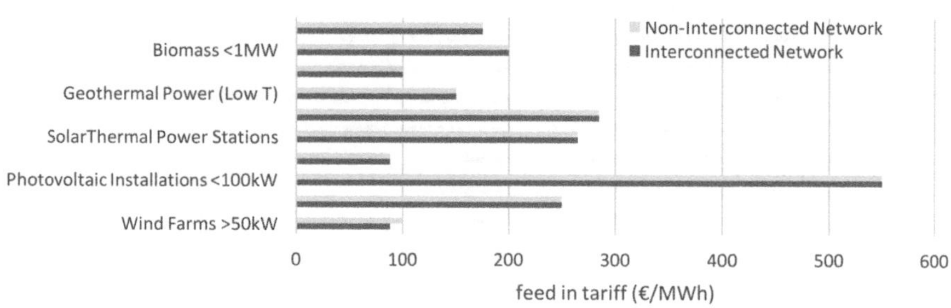

Figure 5.2 FiTs per technology compared, as amended in Act 3851/2010

is based on a day-ahead auction, carried out in order to result in the minimum price per hour covering the system's needs. Such a price is called System Marginal Price (SMP) and equals that of the last bidder to offer their power to the system necessary to cover its needs. Participants in that auction are all non-RES energy producers and are remunerated at the SMP.[3] On the other hand, RES producers, as mentioned earlier, enjoy priority access to the System with no need to participate in the SMP day-ahead auction; their power is supplied into the System irrespective of the outcome of that auction. Since the weighted FiTs exceed historically the weighted SMP, Greece had to establish a financial support mechanism in the form of a RES Special Fund; it comes to no surprise that this was mainly due to the aforementioned high PV FiTs. The purpose of the RES Special Fund is, therefore, to fill the gap between the SMP and the FiT granted to RES producers. The main source of income for this Fund is the Special Levy for the Decrease of Greenhouse Gas Emissions (SLDGE), established by art. 143 of Act 4001/2011. The idea is that energy consumers should bear the cost for avoided emissions.

The evolution of RES incentives: feed-in tariffs rationalized

As expected, high FiTs for PV energy (excessively, as it will be demonstrated below) led to a fast and aggressive evolution of PV capacity in the country and gradually caused an important deficit to the RES Special Fund: a delay on payments to RES producers exceeding 6 months. The decrease of FiTs for PV energy by Act 3851/2010 proved not enough to troubleshoot the issue since a critical part of the PV capacity was already contracted under tariffs provided for in Act 3468/2006. New measures towards the further decrease of FiTs for all RES technologies were inevitable. Act 4254/2014 introduced a weighted decrease of up to 5.4% on average for hydros, 5.6% on average for wind and 29.9% on average for PV (the latter with retroactive effect).

It is revealing to compare the initial span between PV and Wind FiTs and their evolution (2005–2015) as illustrated in Figure 5.3.

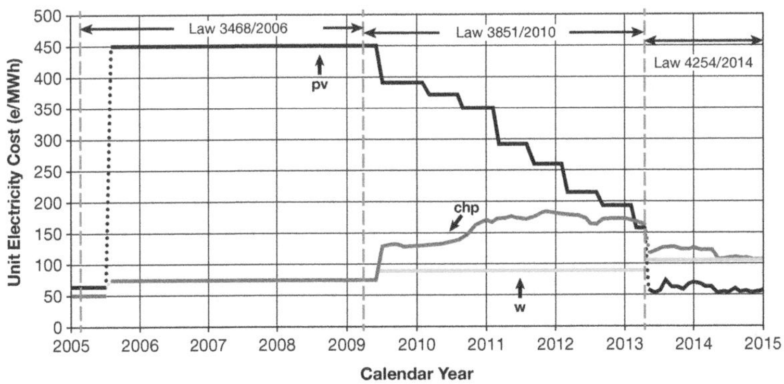

Figure 5.3 FiTs per technology compared, as evolved also under Act 4254/2014

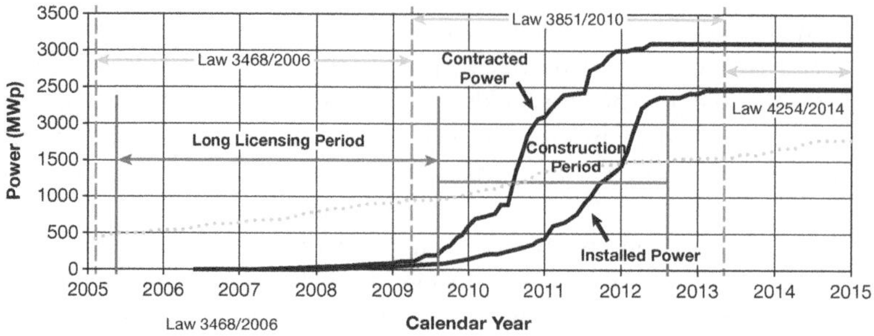

Figure 5.4 RES penetration by technology type compared

Despite strong financial incentives being established in Act 3468/2006, the capacity evolution of PV technology was low throughout the period 2005–2009; this is due to the extremely heavy bureaucratic licencing procedure that caused an average delay of 40 months. The curve of Contracted Power is higher than the curve of Installed Power as an important part of the former was never installed. The result for PV penetration boosted by the FiTs system is depicted in Figure 5.4.

The evolution of RES incentives reloaded: feed-in tariffs finally questioned

High FiTs were finally put into question in the context of recent legislation; more precisely, Act 4414/2016 that introduced a different innovative scheme. This new Act aims at gradually incorporating new RES and Cogeneration of Heat and Power (CHP) units in the Greek electricity market, in a way compatible with the European Guidelines for state aid. It provides for two distinct stages, a temporary and a permanent one; at the first stage, namely for 2016, a supportive mechanism of the RES price is adopted. At the second (permanent) stage – that already started in 2017 – a competitive mechanism of capacity auctions is adopted.

To date, just one Capacity Auction took place on 12 December, 2016 and the auctioned capacity was rather limited, since it only reached 41 MW in total. A total of 13 different new RES units participated, having submitted 481 offers. However, the outcome of this pilot auction is very interesting: the prices awarded in the auction are much lower than those adopted in the FiT scheme as described earlier. Even at a pilot-phase, this new market-oriented scheme effectively reduced the cost of the RES promotion mechanism in Greece. This outcome helps illustrate a small, perhaps reluctant, U-turn in RES incentives at a domestic level. It also offers a more optimistic approach to market instruments instead of regulated tariffs, by highlighting this mechanism as more economically efficient.

Putting the domestic scheme into practice: major negative effects

The favouring of PVs described earlier could not but have considerable effects on the entire RES promotion system. The electricity market has been severely influenced in various ways. Major effects illustrate the non-coherence of the Greek system and the path it has followed towards a near collapse in recent years.

A non-coherent Investment Return Ratio (IRR)

The evolution of PVs IRR starts from 15% in 2006 and increases up to 47% (!) on average by the years 2009–2013. Throughout the latter period, it seems to have oscillated due to various factors: (1) FiTs have been decreasing (on a semi-annual basis) based on Act 3851/2010, and (2) The pace of power installed has decreased. From 2013 PVs' IRR varies between 12% and 21% due to the further decrease of FiTs. The evolution of the wind IRR is considerably different: stability on FiTs led to stable IRR. In recent years, the wind IRR is even higher than PVs' IRR, increasing the attractiveness of wind investments (Figure 5.5).

A non-coherent penetration of RES per technology

Summarizing the effect of FiTs towards the RES penetration to the energy mix (as depicted below) one can say the following:

The deployment of PV energy in Greece may be separated into three different periods: (1) 2006–2009: penetration of PV energy is low despite Act 3468/2006 which offered strong financial incentives. This is primarily due to high administrative costs and red tape causing considerable installation delay. (2) 2010–2013: an aggressive evolution of PV energy is taking place as the PV projects are by now licenced and installed. This is despite the fact that during the same period FiTs are becoming less attractive due to their considerable decrease. (3) 2013–2016: the effect of decreased FiTs is apparent. No considerable power has been added

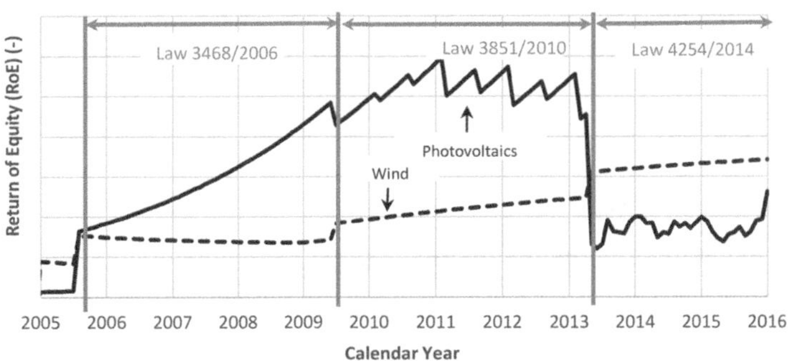

Figure 5.5 IRR per technology, as evolved over the recent years

and the curve of PV evolution is turning to sigmoid. It is stabilized at approximately 2,500 MW installed power of PV technology.

The graph (see Figure 5.6) also presents installed Wind power. By 2005, Wind installed power was already 400 MW due to pre-existing legislation that gave smooth financial incentives. These legislated incentives remained stable by means of FiTs ensuring an annual return on investment of approximately 14%. This caused a stable penetration as it is apparent in the relative curve whose evolution is linear. Other RES technologies have no considerable penetration and equal in total to 250 MW.

Non-coherence in policy support for RES technology, especially for PVs, is not justified on the grounds of technology cost variation. As depicted in the graph (see Figure 5.7), the evolution of installation cost for PVs does not follow the evolution of RES penetration for this specific technology. While installation cost for wind remains almost invariable during the last decade, the evolution of PV is considerably different: starting from 9 Me/MW in 2005, it decreases gradually to less than 1 Me/MW by 2016, mainly due to an immense expansion in PVs industry, alternating the pre-existing equilibrium between offer and demand.

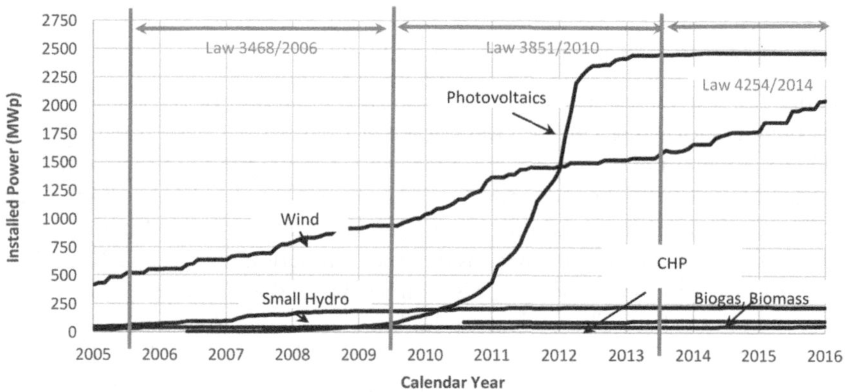

Figure 5.6 RES penetration per technology

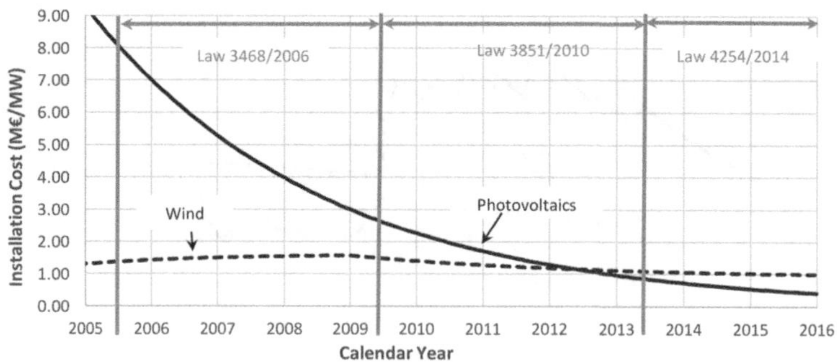

Figure 5.7 RES installation costs per technology

A non-coherent capacity factor per technology

Another criterion to assess the effectiveness of the FiT strategy is the financial correlation between the installed power and the produced energy. As it is depicted in Figure 5.8, 2.467 MW of installed PVs produce 3.58 TWh per year, while 1.751 MW installed Wind farms produce 3.30 TWh annually. It is evident that a much smaller installed Wind power capacity almost equals annual production from PV installations. This is also illustrated in the ratio of contribution to the annual target that Greece has to meet under EU Directive 2009/28 and the Kyoto Protocol: as of 2015, PVs contributed 6.46% and wind farms 5.95%, respectively.

This correlation results in a non-coherent capacity factor for each technology: the PV capacity factor is 16.6, while the equivalent for wind is 22.7, or 40% approximately higher. In other words, despite much higher resources being allocated to PV technology, wind energy production proved to be rather efficient and contributes significantly more in annual production and, subsequently, to the EU 20–20–20 targets. From that point of view, the PV strategy in Greece has proved to be economically inefficient (Table 5.3).

A growing deficit in the RES Special Fund: the inability to pay for FiTs

Another outcome of applying the domestic RES scheme has been a constant and growing deficit in the RES Special Fund (see Figure 5.8). The extent of this deficit depends, first, on the quantity of PV energy absorbed by the energy system, and second, on the amount of the levy actually paid by energy consumers.

As discussed earlier, SLDGE is the core income for the RES Special Fund. Its calculation method is based on the difference between the actual price for RES paid by the system operator and the SMP multiplied by the RES energy. This calculation results in the amount to be paid by consumers so as to cover the

Table 5.3 Measurement of capacity factor and national RES target per RES technology

	Installed power MW	*Load factor %*	*Produced energy TWh/y*	*Target achievement %*
Consumption of electricity			55.5	
Contribution of renewable energy sources				
Photovoltaic installations	2.467	16.6	3.58	6.46
Wind farms	1.751	22.7	3.30	5.95
Small hydroelectric power stations	224	38.0	0.73	1.32
Biogas, biomass	49	50.6	0.21	0.38
Cogeneration of heat and power (CHP)	100	21.3	0.19	0.33
Total of renewable energy sources			8.01	14.4

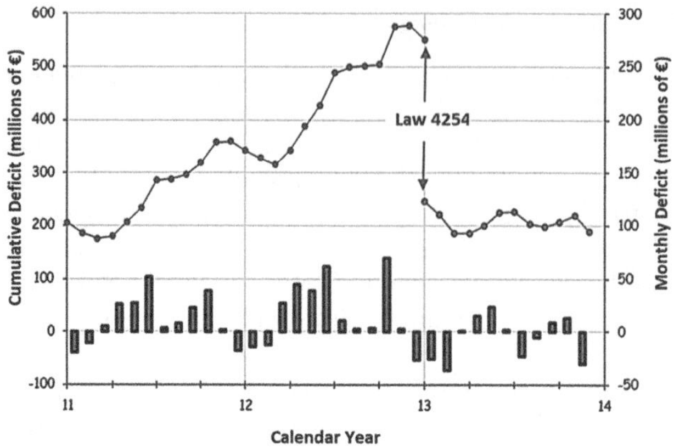

Figure 5.8 Cumulative deficit in the RES Special Fund

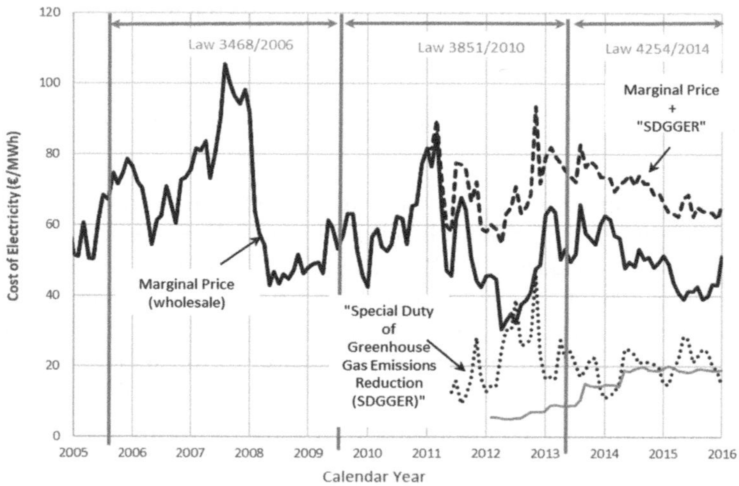

Figure 5.9 Evolution of the SMP and SLDGE

prescribed difference. As both SMP and Actual Price for RES are known, the unknown key factor is the RES energy that the system needs to cover in order to also comply with the national commitment for 2020. There are two methods to calculate the RES energy: either after the period of actual consumption or on a budgetary basis. The Greek operator of electricity market (LAGIE) opted for the second: the operator predicts the RES energy to be consumed over the next semester and the SLDGE calculation is made upon this prediction (Figure 5.9). Nevertheless, such a calculation is extremely difficult and has historically fallen short of reality, causing a deviation between the paid SLDGE and the one that was needed. The said deviation is an additional cause in aggravating the RES Special Fund's deficit.

RES penetration perspective in Greece

The target 20–20–20 and the RES technology penetration evolution to date

The above analysis presented the key factors that influenced the development of renewables in the Greek energy system which are rather financial than regulatory. Regarding the target that Greece has met overall, by the year 2015, the annual electricity production from renewables was approximately 8.2 TWh/year (15% of the national energy product), while the target to meet until 2020 is 11.2 TWh/year (20% of the national energy product). From this perspective, the adopted strategies proved so far to be quite effective (Figure 5.10).

The penetration forecast to 2020

By 2020, the predicted penetration of RES in the Greek energy system will be as depicted in the graph (see Figure 5.11). Photovoltaics will contribute with 6.77% to meet the target of 20%, while wind farms will contribute with 9.67%. The rest of the renewable technologies will contribute with 0.36 minimum (cogeneration) up to 1.43 maximum (small hydro). Regarding PV and wind farms, a considerable difference between their contribution is predicted. Despite the strong financial support to PV technology, wind penetration evolution will bear the major burden to meet the national target (Table 5.4).

The risk for the clean energy target arising from the predicted massive PVs disengagement

Greenhouse Gas Emissions (GHGE) per RES technology are depicted in the next chart. The energy source that emits the least is wind, (10 kgCO_2eq/MWh), while the energy source that emits the most is lignite (100 times higher GHGE

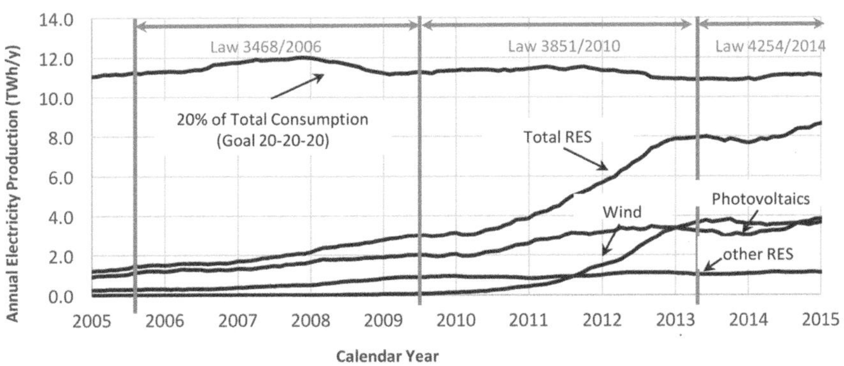

Figure 5.10 Annual electricity production ratio for each RES technology to date

Table 5.4 Annual electricity projection ratio for each RES technology up to 2020

Projected in 2020	*Installed power*	*Load factor*	*Produced energy*	*Target achievement*
	MW	%	TWh/y	%
Consumption of electricity			55.5	
Contribution of renewable energy sources				
Photovoltaic installations	2.580	16.6	3.76	6.77
Wind farms	2.696	22.7	5.37	9.67
Small hydroelectric power stations	239	38.0	0.80	1.43
Biogas, biomass	59	50.6	0.26	0.47
Cogeneration of heat and power (CHP)	108	21.3	0.20	0.36
Total of renewable energy sources			10.4	18.7

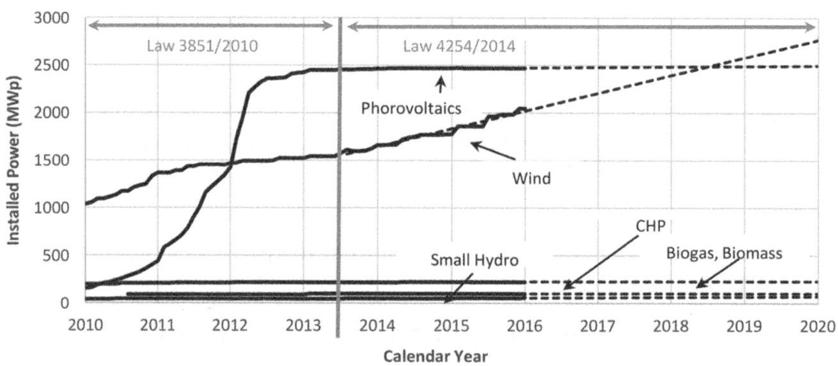

Figure 5.11 Annual installed power projection for each RES technology up to 2020

Table 5.5 GHG emissions per energy source

Electricity generation technology	*Emissions*
	$kgCO_2eq/MWh$
Coal (lignite)	1100
Oil	800
Natural gas	600
Photovoltaics	60
Wind	10

than Wind). PVs emit 60 kgCOeq/MWh, while natural gas emits 10 times higher GHGE than PVs. Emissions are the key reason to promote and support RES penetration, as RES contributes considerably to their reduction. By replacing technologies of high emissions with technologies of low emissions, the effect of avoided GHGE is as presented in Table 5.5: by replacing 1 MWh of energy from lignite with 1 MWh of energy from wind, emissions avoided are 1090 kgCO_2eq per hour.

At this point, it is important to stress that RES installations operate under a 25-year Power Purchase Agreement (PPA) which safeguards their privileged access to the system and specifies the exact FiTs to be paid. Accordingly, the termination of those PPAs will result in the need to restore on time the RES power to be disengaged by another RES, in order to meet the clean energy standards, as they will be adjusted after 2020, at an even higher expected level. Otherwise, what will have been accomplished will be seriously undermined.

However, the expected pace for the disengagement to take place hinges upon the respective pace of penetration: a linear penetration leads to a linear and smooth disengagement, while a steep and uncontrolled penetration results in a sigmoid disengagement, respectively. As depicted below, the prediction of PV and Wind penetration in Greece is the following: wind having a linear evolution for 35 years (2000–2035), its disengagement due to the PPA 25-year effect will be equally linear. Taking into consideration that the wind evolution pace is almost 100 MW per year, the pace of disengagement will be the same. Regarding PVs, the situation is dangerously different: for PVs massively installed with the aforementioned sigmoid evolution that lasted 5 years (from 2007 to 2012), disengagement will last 5 years, respectively. Taking into consideration that the PV evolution pace is almost 500 MW per year, the pace of disengagement will be the same. Nevertheless, the restoration pace needed in this very case is not likely to be achieved, as the PV sector is the one that the FiT legislative reduction hit the most. Consequently, as new power installations are not predicted to take place, a considerable reduction of PV installed power is feared to occur as well. The graph (see Figure 5.12) shows that power amounting to 2 GW will be disengaged from 2032 to 2037. This disengagement pace cannot be restored by any RES technology within a period of 5 years, unless regulatory instruments are put in place in the future to restore the historical irrationalities of the Greek RES scheme.

In terms of restoration of the power to be disengaged, the ideal solution would be to replace the expectedly disengaged RES with the same technology or with a technology that emits even less GHG. The evolution of the avoided emissions is depicted in the graph (see Figure 5.13): for wind, avoided emissions equals 380 ktonCO_2eq/mo, while for PVs avoided emissions equals 150 ktonCO_2eq/mo; the latter will eventually take place between 2032 and 2037. This is yet another effect of the steep and uncontrollable evolution of PVs penetration in Greece.

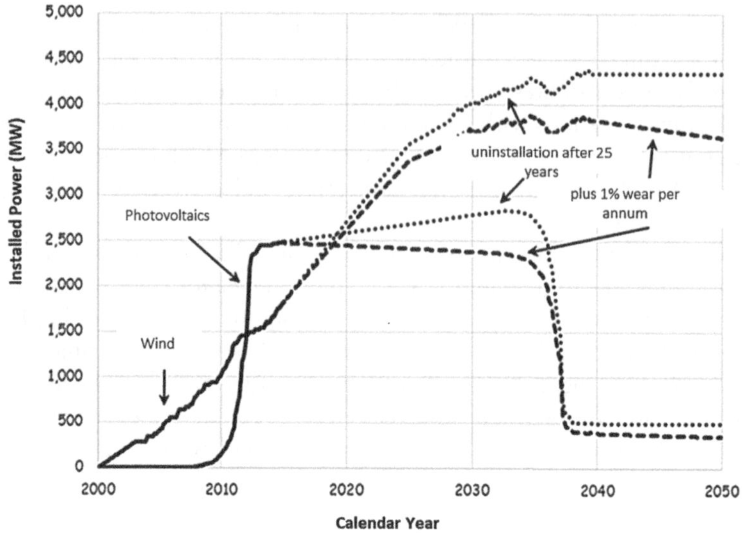

Figure 5.12 Installed power disengagement projection per RES technology

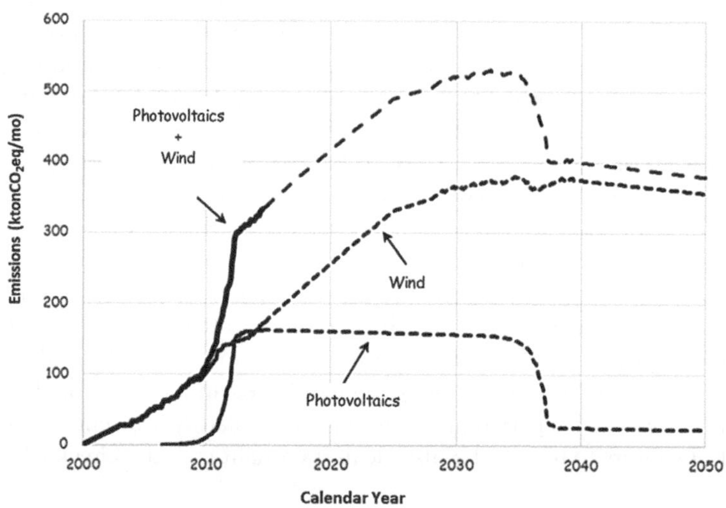

Figure 5.13 Projection of avoided GHGE per RES technology

The Greek case on climate change as a paradigm of an incomplete EU clean energy integration policy

Clean energy ought to form part of a sustainable development policy; however, longing for sustainability with non-viable financial means is by definition contradictory. Moreover, sustainability has an inherent holistic character: if partly undermined, the whole system is in danger to collapse, at least in the long run. The Greek experience is enlightening of the above risk and of those arising from a hasty and incomplete policy. The target 20–20–20, a cornerstone of the EU climate change strategy, is by all means a noble goal; distorting that goal and promoting it with non-coherent instruments jeopardizes its very essence, which is not only to accomplish, but also to maintain a long-lasting swift in energy production. In practice, if Greece followed a slightly different policy regarding RES energy mix, the 20–20–20 target could face a potential but not certain delay. Nevertheless, RES penetration would be more sustainable. Especially for photovoltaics, a different pace of installation that could wait for a period of time for technology costs to drop dramatically, would entail the following: (1) smooth and thus easy to control penetration and disengagement, (2) cost-effective penetration, preventing non-productive rents for PV producers and equipment providers and avoiding an irrational burden for consumers which affects negatively the popularity of climate change policy, and (3) avoidance of significant financial externalities that disperse in the form of an endemic decease to the entire energy sector; timely payments of reasonable FiTs would help reduce exposure for all participants in the energy market and avoid their collapse. At this moment, due to the incurable deficit of the RES Special Fund, all energy players – from the system operator to energy producers, and not only the RES producers – are seriously hampered.

Is all this the fault of the domestic public authorities which opted for a problematic and 'clientelistic' RES scheme? An easy answer would be affirmative (granting excessive FiTs for small PVs was perhaps a vote-fishing instrument) but also myopic. As explained, domestic RES policy is merely an application of the EU regulatory framework under the supervision of the European Commission. *A fortiori*, financial instruments fall under the direct and thorough scrutiny for State aid purposes. Hence, if national FiTs were irrational – which was probably easy to assess at the moment of their establishment – they should have been negatively assessed by EU officials. Moreover, the very existence of the FiT scheme is not of national origin, but an option made available by the EU legislation, legislation that, following art. 106 TFEU, should combine in the best possible way general interest goals with a free market environment.

From that point of view, the original sin for RES penetration can be traced in the lightness by which the European policymakers and legislators treated mechanisms of financial incentives especially for RES, against the fundamental rule according to which subsidies should be the last resort for achieving public interest aims and may be tolerated only insofar as they do not excessively impede free markets. Such rule governs all other areas of the EU energy regulation, such as the financing of Public Service Obligations (PSOs), following the C-280/00, *Altmark* case-law criteria. It was, therefore, a strategic mistake with an ambiguous conformity with EU primary rules to allow through Directive 2001/77/EC an unconditional option between regulated FiTs and Auction mechanisms. The first should be an exceptional complement of the latter, since competition for the market, which is what an auction would serve, should be the rule. Besides, this is what finally happened in Greece with RES, but only after having burdened the whole sector with excessively high subsidies for a quarter of the century through PPAs.

Finally, what has been missing is a more integrated EU policy and regulation for energy and climate change that would limit national discretion on issues as the one examined above. A more aggressive integration would not be contrary to the EU subsidiarity principle since clean energy and climate issues are by definition better off handled at a supranational level. Moreover, such a policy should equally understand energy market and climate change priorities and not treat them as conflicting. For climate change policy to be sustainable – and this is perhaps the most enlightening lesson from the Greek case – market-oriented instruments need to be entrusted to implement targets. All the more so, since the Kyoto Protocol itself and its associated CO_2 emission trading scheme are the most renowned examples of such instruments. Maybe it is not too late yet, even for as a problematic RES scheme as the Greek one, but that is only under the condition that it is treated as an EU problem to be solved through an even more advanced integration.

In its 2030 Energy Strategy, the EU seems to have moved to this direction in a rather bold way: national plans for competitive, secure and sustainable energy, combined with relevant indicators, are established in order to ensure a greater coherence, further market integration and certainty to investors. At the same time, it is acknowledged that higher shares of RES are not in themselves enough to meet the 2030 energy targets; systematic monitoring at an EU level should also be in place to achieve a more efficient energy system throughout the EU.[4] These steps aim at ensuring a more thorough and consistent scrutiny from the European Commission on national measures, including incentives to promote RES. They promise a deeper integration, in line with the EU subsidiarity principle. However, their introduction is only the beginning; application by the European authorities needs to be safeguarded and be strict and coherent among all member states. Guidance by the Commission as envisaged in the 2030 Energy Strategy is probably the most crucial tool in this new EU energy policy: such guidance needs to guarantee a streamlining of national plans, especially with regard to financial incentives, by establishing economic efficiency criteria for national authorities to meet when designing and operating domestic RES promotion schemes. Furthermore, Commission supervision that proved to be insufficient until today can benefit considerably national authorities especially in the context of the iterative process provided and ensure that a new Greek case of inefficient RES introduction is avoided.

Notes

1 Directive 2001/77/EC of the European Parliament and Council of September 27, 2001, on the promotion of electricity produced from renewable energy sources in the internal electricity market (OJ L 283/27.10.2001).
2 Directive 2009/28/EC.
3 Although some extra mechanisms are put in place for other reasons, such as the need to establish the desired power in the system, it is beyond the scope of the present analysis to go into further detail as this does not alter the conclusions drawn herein.
4 See Communication from the Commission to the European Parliament, the Council, the European Economic and Social Committee and the Committee of Regions – A policy framework for climate and energy in the period from 2020 to 2030, COM(2014) 15 final.

Bibliography

Alberth, S. (2008). "Forecasting Technology Costs Via The Experience Curve – Myth Or Magic?" *Science Direct* 75, 952–983.

Agency, I.E. (2015). *World Energy Outlook 2015/Intergovernal Panel on Climate Change 2014/5th Asssesment Synthesis*. Paris, France.

Arpino F., Cortellessa, G. and Frattolillo, A. (2015). "Experimental and Numerical Assessment of Photovoltaic Collectors Performance Dependence of Frame Size and Installation Technique." *Solar Energy* 118, 7–19.

Battle, C., Perez-Arriaga, I.J. and Zambrano-Barragan, P. (2012). "Regulatory Design for Res-E Support Mechanisms: Learning Curves, Market Structure and Burden-Sharing." *Energy Policy* 41, 212–220.

Braden, J.B., Folmer, H. and Ulen, T.S. (n.d.). *Environmental Policy with Political and Economic Integration. The European Union and the United States*. Edward Elgar, Northampton, MA.

Buonocuore, J.J., Luckow, P., Norris, G., Spengler, J.D., Blewald, B., Fischer, J. and Levy, J.L. (2016). "Health and Climate Benefits of Different Energy – Efficiency and Renewable Energy Choices." *National Climate Change* 6, 100–105.

Castro Santos, L. and Diaz-Casas, V. (2015). "Sensitivity Analysis of Floating Offshore Wind Farms." *Energy Conversion and Management* 101, 271–277.

Chichilnisky, C. (1997). "What Is Sustainable Development." *Land Economics* 73, 4, 467–491.

Ederer, N. (2015). "Evaluating Capital and Operating Cost Efficiency of Offshore Wind Farms: A DEA Approach." *Renewable and Sustainable Energy Reviews* 42, 1034–1046.

Farfan, J. and Breyer, C. (2017). "Structural Changes of Global Power Generation Capacity Towards Sustainability and the Risk of Stranded Investments Supported by a Sustainability Indicator." *Journal of Cleaner Production* 141, 370–384.

Fee, D. (1987). *Third Party Financing for Energy Savings*. Jessica Kingsley Publishers, London.

Giannini, E., Moropoulou, A., Maroulis, Z. and Siouti, G. (2015). "Penetration of Photovolatics in Greece." *Energies* 8, 6497–6508.

Kaldellis, J.K. (2002). "An Integrated Time – Depending Feasibility Analysis Model of Wind Energy Applications in Greece." *Energy Policy* 30, 267–280.

Kaldellis, J.K. (2011). "Critical Evaluation of Financial Supporting Schemes for Wind-Based Projects: Case Study Greece." *Energy Policy* 39, 2490–2500.

Kaldellis, J.K. and Zafirakis, D. (2007). "Present Situation and Future Prospects of Electricity Generation in Aegean Archipelago Islands." *Energy Policy* 35, 4623–4639.

Kates, R.W., With National Research Council, Board on Sustainable Development. (1999). *Our Common Journey a Transition Toward Sustainability*. National Academy Press, Washington, DC.

Katsaprakakis, D. (2012). "A Review of the Environmental and Human Impacts from Wind Parks. A Case Study for the Prefecture of Lasithi, Crete." *Renewable and Sustainable Energy Reviews* 16, 2850–2863.

Mir-Artigues, P., Gerda, E. and Del Rio, P. (2015). "Analyzing the Impact of Cost Containment Mechanisms on the Profitability of Solar Pv Plants in Spain." *Renewable and Sustainable Energy Review* 46, 166–177.

Tsaltas, G. and Katsibardis, C.G. (2004). *Sustainable Development and Environment – The European and National Perspective*. I. Sideris Publishing House, Athens, Greece.

Tsaltas, G.I. and Platias, C. (2010). *The European Union and the Environment*. I. Sideris Publishing House, Athens, Greece.

Verde, S.F. and Pazienza, M.G. (2016). "Energy and Climate Hand in Hand: Financing Res-E Support with Carbon Revenues." *Energy Policy* 88, 234–244.

6 The sustainable energy transition through international and EU law

Stuart Bruce[1]

Introduction

The negative environmental, social and economic effects of a fossil fuel-based society, derived from a time of inadequate scientific knowledge and protective legal frameworks, are increasingly prevalent around the world. Among the responses and approaches to addressing these global challenges is the development and promulgation of sustainable energy. The European Union ('EU') was an early adopter, and remains a world leader, in the development of sustainable energy laws and policies designed to usher in a clean, sustainable and modern society.

As this book highlights, the global transition to sustainable energy generation and use is well underway and has gained pace significantly in recent years. The initial shift towards sustainable energy was encouraged by the moral imperative to mitigate risks of catastrophic climate change. But as the economic and technical viability of sustainable energy has improved – particularly when coupled with battery storage – it is increasingly seen as a sensible option for consumers, businesses and governments, independently of considerations of morality.

Public international law abounds with legal instruments promoting environmental law. Much of this has an indirect effect on sustainable energy norms and policies. The EU has taken this a step further through a host of legal instruments that address sustainable energy issues directly, and not just incidentally to the wider issue of environmental protection. Legal frameworks and regulations are the EU's principal means of influencing investment interest in sustainable energy and setting clean energy targets. Together, these help to address the trilemma of decoupling economic and emissions growth, while also strengthening energy independence – the benefits of which have been endorsed by the Group of Twenty ('G20') (G20 2017, p. 4).

The EU's sophisticated framework offers an example to other regional organizations and nations seeking to promote the environmental and economic benefits of sustainable energy. The legal drive to energy efficiency has benefited the EU through both savings on more costly energy sources and encouraging investment in new energy technologies, with European investment in energy efficiency rising by 10% between 2015 and 2016 alone (IEA 2017, p. 91). Completing the solution to the trilemma, sustainable energy policy has also strengthened EU energy security by reducing peak daily gas demand and weaning Europe off imports. In Germany alone, the energy intensity of residential space heating improved by 35% between 2000 and 2012 (IEA 2017, p. 32).

This chapter presents a survey of the public international law and EU sustainable energy regimes, showing the role that legal reform has played in both setting and meeting sustainable energy agendas in Europe. Beyond treaties and protocols, regional energy policy has a 'soft law' effect which develops norms and influences investor and consumer behaviour. Thus, EU sustainable energy law has both direct and indirect effects: for example, through a combination of mandated targets and incentives that not only set down legal norms, but change how people think about the very issues that sustainable energy law addresses. EU law, therefore, has the sign-posting effect of public international law, but backs this up by mandating real clean energy targets and setting down prescriptive regulations. The range of actors involved, from legislators to investors, demonstrates the flexibility and variety that characterizes the European legal regime.

Part II of this chapter considers the largely indirect influence of public international law on sustainable energy generation and use. Part III considers the highly influential EU legal system, which has changed the international community's conception of renewables and energy efficiency (see, e.g., Kochenov and Amtenbrink 2014). Part IV remarks on possible directions for future research.

The international legal order relating to sustainable energy

The international legal order relating to sustainable energy is complex, multi-layered and under construction. It is not a self-contained regime, but consists of an interplay between international environmental law norms, non-binding international policy instruments or 'soft laws', directly and indirectly applicable binding treaty obligations and, increasingly, international actors and institutions.

Most treaty-based regimes do not impose binding obligations with specific targets for the use and generation of sustainable energy. Instead, they largely attempt to create structures and incentives to encourage behaviours and choices that favour sustainable energy technologies and policies. By comparison, non-binding instruments in the form of declarations, resolutions and guidelines have proliferated and played a key role in the development of international (and national) sustainable energy law and policy. In turn, policy objectives have been advanced by a broad array of actors. The EU has been particularly influential in shaping many of these legal instruments, directly through intergovernmental negotiations and indirectly through its external relations (Boute 2013).

Rules, principles and norms

The core rules and principles of international environmental law, though not directly applicable to the energy sector, have influenced the evolution of energy policy and promoted sustainable energy (Oniemola 2013, p. 242). While certain of these principles, such as the duty not to cause significant transboundary harm,[2] have crystallized into customary international law, the legal status of most principles relevant to energy activities, including precaution, polluter pays and sustainable development, remain contested (Bruce 2013, pp. 29–30; Oniemola 2013, p. 244). These principles have shaped international, regional and national policy in favour of sustainable energy and will likely continue to influence developments.

For example, to implement the 'no harm principle' states may choose to discourage fossil fuel-based energy sources in preference for non-polluting sources, thereby furthering the use of renewable energy and energy efficiency (Bradbrook 2008, p. 114). More widely, the duty might extend to avoiding the use of energy facilities that contribute to global warming; but as with other international environmental law principles, this position is not sufficiently developed in law (id.).

Certain principles and other norms impacting energy are enshrined in soft law instruments, non-binding declarations and resolutions adopted at political conferences and summits, which provide quicker and more flexible responses to coordinated intergovernmental action than treaties (Redgwell 2006, p. 97). Of the mega-conferences (see generally Najam and Cleveland 2005), the *1992 Rio Declaration*[3] constituted a cornerstone in the growth of sustainable development policy, promoting sustainable energy by encouraging reduction and elimination of 'unsustainable patterns of production and consumption'.[4] The contemporaneous *Agenda 21*[5] requested governments to promote renewable energy research and technology transfer, including a review of the energy supply mix, and to use 'environmentally sound energy sources' such as sustainable energy technologies to address environmental concerns (Oniemola 2013, p. 247). This program was further bolstered by the UN General Assembly, which in 1997 called upon states to increase renewable energy power generation.[6]

2002 marked a turning point for international sustainable energy cooperation. Adopted at the World Summit on Sustainable Development, the *Johannesburg Plan of Implementation*[7] highlighted the connections between energy security, climate change and sustainable development and provided normative force for delinking economic growth and environmental degradation (Bradbrook and Wahnshafft 2005, p. 195). Its encouragement of states to increase urgently the global share of renewable energy sources[8] was subsequently codified in the 2011 UN Sustainable Energy for All initiative, which seeks to transform the global energy sector through a doubling of renewables and energy efficiency and the provision of universal access to electricity.[9] Those goals have been transposed into Goal 7 of the Sustainable Development Goals, to 'ensure access to affordable, reliable, sustainable and modern energy for all'.

Treaty law and renewable energy

There is yet no specific global treaty governing renewable energy. None of the treaties touching on sustainable energy issues establish binding energy generation and consumption obligations. Rather, they have been designed to incentivize decentralized sustainable energy policy choices and investment through the creation of innovative mechanisms and support measures.

Climate change regime

The international climate change regime consists of a constitutive framework treaty with guiding principles, a regulatory protocol with operative targets and mechanisms to reduce greenhouse gas emissions, intricate institutional decisions to advance long-term parallel programmes, and, most recently, an agreement to encourage transparency of, and ambition in, domestic climate-related actions. Collectively, these mechanisms help to create conditions and support structures that foster sustainable energy transition.

UNITED NATIONS FRAMEWORK CONVENTION ON CLIMATE CHANGE

The UNFCCC entered into force in 1994. Its 'ultimate goal' is to stabilize greenhouse gas concentrations in the atmosphere – which continues to increase from use of conventional fossil-fuel energies – at a level that averts catastrophic climate change.[10] However, it imposes no meaningful obligations on state parties (which it differentiates according to their development status) relating to emissions reductions or energy generation and use, and no legal obligations or policy prescriptions related to sustainable energy.

The central obligations require each party to establish national inventories of greenhouse gas emissions and develop national and/or regional measures to mitigate climate change, guided by principles of intergenerational equity, precaution and sustainable development.[11] Although renewable energy is not expressly mentioned, and energy efficiency is referenced in the preamble only as a means to encourage environmentally sound 'sustainable social and economic development', parties are required to promote and cooperate in the development and transfer of technologies that control, prevent or reduce greenhouse gas emissions in industry and in energy, transport and waste sectors.[12] Collectively, these obligations indirectly promote sustainable energy.

UNFCCC KYOTO PROTOCOL

The Kyoto Protocol, adopted in 1997 and entered into force in 2005, imposes binding emissions reductions targets on developed states, which must be achieved during the 2008–2012 and extended 2013–2020 compliance periods. Parties are obliged to implement domestic policies to achieve the emissions reductions goals. The text provides a non-exhaustive list of eight non-binding policy options that parties may adopt domestically, including energy efficiency and the use of new and renewable forms of energy,[13] though parties retain discretion over their policy and technology choices.

To achieve the emissions reduction obligations, three innovative market mechanisms were established: the Clean Development Mechanism ('CDM'), Joint Implementation ('JI') and the Emissions Trading Scheme ('ETS'), all of which were intended to direct investments towards low carbon energy technologies and other forms of emissions reductions.[14] The emissions abated from CDM and JI projects create credits that can count towards the parties' international emissions reductions obligations.[15] The effect of this is that CDM and JI make low-emission projects competitive against cheaper but higher-emission projects, enable investments in low-carbon projects and technology transfer to developing countries.[16] In practice, eligible CDM projects have generated considerable investments in sustainable energy, and abated tens of millions of tonnes of emissions.

The Kyoto Protocol ETS has never been established, but states and regions have created their own ETSs founded on similar philosophies. An ETS incentivizes development of innovative low-carbon technologies by capping emissions and creating a market for emission reduction units (Netto and Schmidt 2005, Part 7), which may satisfy domestic and/or international obligations.

SUBSEQUENT DEVELOPMENTS AND THE PARIS AGREEMENT

Beyond treaty texts, the climate change regime has developed a complex matrix of actions and obligations. Parties have negotiated, among other things, instruments related to forestry, transportation, industrial, food and agriculture, housing and infrastructure sectors; adaptation for coastal, desert, rural agricultural and other vulnerable people; and support for new carbon markets and climate finance.[17]

Various technology and finance mechanisms have also been developed to assist with technology transfer and climate financing for developing state parties. For example, a Green Climate Fund was established to secure up to $100 billion per year in new financing for climate-related projects, and a Technology Mechanism was introduced to facilitate the process of clean technology knowledge sharing, particularly for developing countries.[18]

Most recently, the Paris Agreement was adopted in December 2015 and entered into force early in November 2016. It establishes the parameters and governance structures for the global climate regime post-2020 and represents unprecedented universal cooperation to tackle climate change and to 'change the course of the global economy' (Figueres 2017, v). It sets a long-term goal for global temperature increases to be 'well-below 2C above pre-industrial levels', an aspiration target of 1.5C, and effectively requires global peaking of greenhouse gas emissions then net global decarbonization in the second half of this century.[19]

The parties agree to 'mak[e] finance flows consistent with a pathway towards low greenhouse gas emissions and climate-resilient development',[20] thereby directing future funds to sustainable energy projects. Parties also acknowledge the importance of securing 'universal access to sustainable energy in developing countries, particularly in Africa'.[21] These objectives will be advanced through rolling five-year Nationally Determined Contributions ('NDCs'), which are voluntary actions, targets and plans that states commit to undertake post-2020. Sustainable energy policies and projects will be central to the attainment of those NDCs, including the EU's pledge to reduce its greenhouse gas emissions by 40% from 1990 levels by 2030, and to improve the ETS so that it is fit for purpose (European Commission, 2030 Climate and Energy Framework). But despite clear signs of progress, the most recent studies suggest that even the progressive EU will fall short of its NDC target for 2030 under existing policies, stressing a fortiori the need for an effective regional sustainable energy regime (UNEP 2017).

Other sectoral, geographic or particular regimes

Other treaty regimes do not directly or indirectly incentivize sustainable energy projects, but are relevant to their regulation and governance. Selected treaties are briefly noted below. The trade regime is discussed elsewhere in this book.

The United Nations Convention on the Law of the Sea is particularly significant for offshore renewable energy production. The treaty establishes coastal state jurisdiction over the construction and operation of offshore wind turbines and marine renewable energy generation within that state's territorial seas, exclusive economic zone or on its continental shelf,[22] subject only to rights and obligations of other states relating to innocent passage, freedom of navigation and the laying of submarine cables and pipelines.[23] It also provides for the sovereign rights of coastal states to exploit the living and non-living resources of the water column, seabed and subsoil, including ocean energy.[24] Regional seas agreements dealing with particular issues are also applicable.

In relation to international public planning laws, the Aarhus Convention mandates access to information, public participation in decision-making and access to justice in environmental matters, and is applicable to larger sustainable energy projects.[25] Similarly, the Espoo Convention requires states to undertake an environmental impact assessment, notify and consult with states on major projects that may have an adverse transboundary environmental impact.[26] Its protocol requires parties to integrate a strategic assessment of the environmental consequences of their plans and programmes prior to impact assessments.[27] These instruments have particular geographic application in Europe.

Nature protection treaties are relevant to locating renewable energy projects, maintaining biological diversity,[28] and conserving migratory species of wild animals[29] and important wetlands.[30] Relatedly, one of the few treaties expressly addressing energy in relation to geographies of natural beauty is the energy protocol to the Alpine Convention, which sets out obligations regarding sustainable development, energy efficiency, energy conservation and renewable energy production for Germany, France, Austria, Lichtenstein, Monaco, Slovenia and Switzerland.

In international economic law, multilateral and bilateral investment treaties protect against sovereign risk by affording foreign investors minimum standards of protection for their investments in covered projects, including sustainable energy projects (see generally Dupuy and Viñuales 2013). Disputes would usually be determined by an international tribunal, itself a valuable right.

Treaty law and energy efficiency

The only multilateral treaties directly related to the energy sector, and energy efficiency, are the 1994 Energy Charter Treaty ('ECT') and its Protocol. This section briefly considers the origins, content and evolution of these instruments, which are largely limited to Europe and have been subject to multiple arbitrations in recent years.

Energy Charter regime

ENERGY CHARTER TREATY

The ECT was born from the post-Cold War political process, the Energy Charter, in which states agreed to a legal and institutional framework to ensure integrated and efficient operation of energy markets in Europe without interruption. The ECT establishes a non-discriminatory, open international energy market by requiring mutual cooperation for free trade in energy materials and products, freedom of energy transit through pipelines and grids, promotion and protection of energy investments and the option of international arbitration for energy disputes (see, e.g., Reed and Martinez 2008; Happ 2002).

The ECT attempts to address sustainable energy through environmental protection and an energy efficiency protocol. States are required to minimize harmful environmental impacts from energy operations within their territory by taking into account various international environmental obligations. The text also imposes approximations of principles from the *Rio Declaration*, including polluter-pays and precautionary measures.[31] Additionally, states are explicitly required, among other things, to 'have particular regard' – a weak form of legal obligation – to improving energy efficiency and developing renewable energy sources.[32] There is also an indirect obligation to take account of externalities.[33]

ECT PROTOCOL ON ENERGY EFFICIENCY

The Energy Charter Protocol on Energy Efficiency and Related Environmental Aspects ('PEEREA') entered into force in 1998. PEEREA recognizes that 'improvements in energy efficiency' diminish negative environmental impacts,[34] and seeks the 'promotion of energy efficiency policies consistent with sustainable development'[35] and 'fostering cooperation in the field of energy efficiency'.[36] It also creates framework conditions to 'induce producers and consumers to use energy as economically, efficiently and environmentally soundly as possible, particularly through the organization of efficient energy markets and a fuller reflection of environmental costs and benefits'.[37] Like the ECT, PEEREA has been criticized for its vague and non-binding wording (Bradbrook 2008, p. 118), using 'largely hortatory terms', and lacking enforcement machinery (Redgwell 2016, p. 99). Practically, the PEEREA has facilitated cooperation and knowledge exchange through Working Group dialogues with states developing energy legislation, analytical studies and voluntary country reviews.

INTERNATIONAL ENERGY CHARTER

In recent years, the Energy Charter political process undertook a modernization process resulting in, among other things, the adoption in 2015 of the non-binding International Energy Charter ('IEC'). The IEC is open to global membership, and aims to strengthen energy cooperation between signatory states. It is 'desirous of sustainable energy development', 'maximizing the efficient production, conversion, transport, distribution and use of energy',[38] and recognizes that energy efficiency and renewables 'enhance energy security and contribute to sustainable economic growth'.[39]

Significantly, the IEC's third objective is energy efficiency and environmental protection. It aims to create 'mechanisms and conditions for using energy as economically and efficiently as possible, including, as appropriate, regulatory and market based instruments', and increase the sustainable energy mix and clean energy sources. The latter is conditioned upon minimizing negative environmental consequences from those activities in 'cost-effective ways'.[40] While the IEC's sustainable energy aspirations are articulated in greater detail than in the ECT and PEEREA, they are non-binding. It is, therefore, another normative declaration supporting sustainable energy and greater levels of state practice.

International energy actors

There is no single international organization with a mandate to regulate all energy matters. Much like the legal frameworks, international governance of sustainable energy consists of a patchwork of disparate international actors with various mandates and special interests. This patchwork encompasses energy organizations, UN agencies, political forums, development banks, think tanks and NGOs, which, collectively, form a network of actors that shape, develop and constitute international sustainable energy governance (Leal-Arcas and Minas 2016, p. 5). The key actors are noted below.

International Renewable Energy Agency

The opening of IRENA in 2011 marked a watershed for global governance of, and political commitment to, renewable energy. IRENA, now with 156 members and 24 states in the process of accession, is the only intergovernmental organization with a dedicated mandate to 'promote the widespread and increased adoption and the sustainable use of all forms of renewable energy'.[41] IRENA plays a pivotal role in enabling the transfer of renewable energy knowledge and technology, capacity building, concessional finance, and organizing systematic activities that comprise states, intergovernmental and non-governmental organizations and private actors to encourage the use of renewable energies (Bruce and Stephenson 2016, p. 10). It can only provide recommendations, however. It cannot create international legal obligations regarding renewable energy.[42] Yet, through its products and programs such as the handbook on nationally appropriate mitigation actions, the Global Atlas for Renewable Energy Resources, the Renewable Energy Roadmap, the Abu Dhabi Fund for Development and reports on solar PV markets, electricity storage and state-level assistance, IRENA has quickly demonstrated its influence.

United Nations bodies and development banks

Energy is not mentioned in the UN Charter, but falls within the UN's broad and general competence. The UN's approach and attention to sustainable energy has varied. Since the 1981 Nairobi Conference on New and Renewable Sources of Energy, various programmes and bodies within the UN system have related to sustainable energy. Most prominently, the Commission on Sustainable Development dedicated its 9th and 15th sessions to energy, providing considerable knowledge exchange and political dialogue, but demonstrated the divergent views among states (see Hodas 2011). Today, UNDP, UNEP, UNIDO, FAO, UNESCO, WMO, UN-Habitat and many other bodies undertake initiatives and activities to advance sustainable energy. In 2004, UN-Energy was established as a mechanism for inter-agency collaboration across programmes, though its impact is queried.

International financial institutions, including multilateral development banks, increasingly play a crucial role for long-term and higher-risk funding for the transition to a low-carbon economy, particularly in sustainable development projects. The International Finance Corporation and all major development banks have established programmes and financing for sustainable energy projects. As of July 2016, over $30 billion was committed towards achieving the targets of SDG 7 alone.[43] Relatedly, the Global Green Growth Initiative is a new institution aimed at accelerating socially inclusive and environmentally sustainable economic growth in developing and emerging economies.

International Energy Agency

The IEA was formed in response to the 1973–1974 oil crisis to coordinate a collective response to security issues among oil-importing countries. In 2005, the IEA's mandate expanded to consider low-emissions energy sources, including renewable energy and energy efficiency.[44] It has since released special reports on climate change, energy and low-carbon technologies, enabled cooperation on clean energy policy among members, and measures and reports on sustainable energy developments and policies through its databases. The IEA's annual flagship publication, the World Energy Outlook, is a preeminent source of global energy data and analysis.

Political forums

The many co-benefits of sustainable energy were first recognized at the international political level in 2005, during the Group of Eight Gleneagles Summit, where the IEA's mandate was expanded. Thereafter, the political forums, including the Group of Seven ('G7') and the G20, have sent policy signals to the market by repeatedly emphasizing the benefits of sustainable energy to improve global energy security cooperation, diversification of energy mix, substitution of fossil fuels and reductions of greenhouse gases. From 2014, the G20 has been committed to sustainable development through cost-effective clean energy and has acknowledged 'the priority' of energy efficiency to strengthen energy security and independence by creating the first international policy agreement (other than PEEREA) on energy efficiency. In 2015, the G7 agreed to decarbonize the global economy during this century (G7 2015, p. 12). While the political signals are encouraging, further political leadership will be necessary for a successful global transition to renewables (Leal-Arcas and Minas 2016, pp. 30, 35–36, 46).

Bilateral political forums are also increasingly serving as a mechanism for advancing climate change and sustainable energy goals. For example, on 6 October 2017, the EU and India released a non-binding joint statement on cooperation to increase the energy efficiency of products, industrial processes and buildings and to further cooperate on the smart integration of renewable energy into the domestic electricity system, 'including the enabling policy and regulatory aspects' ([15(d)]–[15(e)]). How such models will advance the stated goals and interplay with traditional multilateral relations will provide fertile ground for future examination. States are also deploying significant unilateral domestic measures to transition to low-carbon economies, mindful of their economic, social and environmental co-benefits.

NGOs, think tanks and multi-stakeholder partnerships

The current international dialogue and awareness of sustainable energy would likely not be as buoyant were it not for NGOs, think tanks and various forms of multi-stakeholder partnerships keeping it on the agenda for decades. Organizations such as WWF, the IUCN and IISD were vocal advocates of sustainable energy during the 1990s. REEEP and Ren21C arose out of intergovernmental conferences in the early 2000s specifically to promote clean energy projects and markets in emerging economies, to monitor national policy developments and develop highly regarded publications such as the Global Status Report on renewables, and, since 2017, on energy efficiency. More recently, high-net worth individuals and modern-day industrialists have established foundations and agencies promoting sustainable energy transition.

European Union law and regulation on sustainable energy

The EU's legal frameworks and policies on sustainable energy are among the most progressive and influential. To support its mission for economy-wide decarbonization by 2050 (European Commission, 2050 low-carbon economy), it has designed and implemented ambitious policies, legally binding targets and support policies to influence sustainable energy policy choices.

As part of its 2020 climate and energy package, the EU must reduce greenhouse gas emissions by 20% on 1990 levels, ensure 20% of final energy consumption is generated from renewables and improve energy efficiency by 20%, which are objectives to be advanced through differentiated national targets for member states. The EU is on track to achieve and surpass these goals (see, e.g., European Commission, Fact Sheet: Renewables: Europe on track to reach its 20% target by 2020). In October 2014, the European Commission increased the aspirations of its 2030 climate and energy framework, proposing that it reduces emissions by 40% by 2030, and the renewable energy and energy efficiency goals each be increased to 27% (European Commission, 2030 climate and energy framework). The regulatory model of this package sets EU-level targets, without binding national targets, encouraging differentiated activities for each member state's circumstances.

In November 2016, as part of its Clean Energy for All Europeans package, the European Commission proposed increasing the energy efficiency goal to 30% by 2030 and observes that 50% of European electricity should by then be renewable (European Commission, 30 November, 2016b). Additionally, the EU's 2050 Energy Strategy sets objectives for nuclear energy and carbon capture and storage as well as renewables and energy efficiency.[45]

The legal frameworks for these goals are outlined below.

Renewable energy

The principal mechanism by which renewable energy sources are regulated in Europe is through the Renewables Directive, adopted in 2001 and amended in 2003 and 2009, the purpose of which is to establish 'a common framework for the promotion of renewable energy from renewable sources'.[46]

A significant innovation is the Directive's mandatory (and differentiated) national targets for each member state's gross final consumption of renewable energy by 2020, and for each to ensure by then that 10% of the energy used in transport is from renewable sources. Biofuels and bioliquids are, therefore, instrumental in assisting the achievement of their targets. The Directive sets out biofuels sustainability criteria for all biofuels produced or consumed in the EU.[47] Cumulatively, the targets will enable the EU to achieve its 20% renewable energy target in the total energy consumption mix by 2020.[48] While the Directive mandates quantitative outcomes, it largely preserves sovereignty choice over renewable energy technologies, policies and sectors, as informed by national circumstances and geological endowments.

Member states may utilize cooperative measures with third states and EU support schemes, including instruments and mechanisms applied by an individual or group of member states to promote renewable energy by reducing its cost, increasing its price or increasing the volume purchased. Examples include investment aid, tax exemptions or reductions, tax refunds, and feed-in tariffs and premium payments.[49] Information about the support schemes is to be provided to such actors as consumers, builders, architects and suppliers, and information about the net benefits of renewables are to be provided by suppliers or administrative authorities.[50] Cooperative measures include joint support schemes and renewable energy sharing among member states,[51] and joint projects among member states and third states to produce electricity, heating and cooling from renewable energy sources.[52]

To help realize their targets and implement the Directive, each EU member state is required to adopt national renewable energy action plans and be able to guarantee the origins of the renewable energy supplied.[53] The action plans are to set out national targets for renewable energy across sectors, the measures to be taken to advance those targets and individual or joint support schemes, renewable energy sharing and biomass policies.[54]

In addition to targets and policies, member states are obliged to develop transmission and distribution grid infrastructure, intelligent networks, storage facilities and the electricity system, as well as interconnection between member states and between member states and third countries.[55] In this respect, it also imposes obligations for transmission system operators and distribution system operators. Where the member state establishes national rules concerning the authorization, certification and licensing procedures applicable to plants and transmission and distribution network infrastructures for the production of electricity, heating or cooling from renewable energy sources, those rules are to be proportionate, necessary and transparent.[56]

The Directive thus seeks to increase energy security and independence, and promote investor stability and certainty to encouraging further investments into renewable energies, while also reducing greenhouse gas emissions (Leal-Arcas and Minas 2016, p. 39). The same objectives are advanced through energy efficiencies.

Energy efficiency

The principal mechanism by which supply and demand side energy efficiency is regulated in Europe is the 2012 Energy Efficiency Directive. The Directive's purposes are to improve supply and demand side energy efficiency in the EU by 20% by 2020. To do this, it establishes a 'common framework of measures for the promotion of energy efficiency' in the EU and provides rules to remove barriers in the energy market and overcome market failures impeding efficient supply and use of energy.[57]

Unlike the Renewables Directive, the Energy Efficiency Directive does not mandate national targets. It delegates to member states the obligation to set their own indicative targets. These may be based on primary or final energy consumption, primary or final energy savings, or energy intensity, taking into consideration national circumstances affecting primary energy consumption including cost-effectiveness, GDP, energy imports and exports and the development of all sources of renewable energies, including nuclear and carbon capture and storage.[58] All member states must use energy more efficiently at all stages from production to final consumption.

On the energy supply side, member states are to promote efficiency in heating and cooling, and improve energy transformation, transmission and distribution. Heating and cooling is to be improved by undertaking and reporting to the European Commission an assessment of the potential for high-efficiency cogeneration and efficient district heating and cooling.[59] When operating gas and electricity infrastructure under other directives, national energy regulators are to pay due regard to energy efficiency, and when developing network tariffs and regulations to take into account the costs and benefits of each measure and provide incentives for grid operators to deploy smart grids.[60]

On the energy demand side, the obligations are extensive and pertain to public and other buildings, energy management and energy efficiency obligation schemes ('EEOS'). Member states are to purchase energy efficient buildings, products and services, establish a long-term strategy for mobilizing investment to renovate the national stock of private and public residential and commercial buildings, and carry out energy efficiency renovations to improve heating, cooling and minimum energy performance standards on at least 3% (by floor area) of the buildings their governments own and occupy.[61] Such information is to be reported to the European Commission as part of the National Energy Efficiency Action Plans mandated under an earlier end-use energy efficiency directive.

In relation to energy management, consumers of electricity, natural gas, district heating and cooling and domestic hot water are to be provided with accurate individual metering, where technically and financially reasonable; and free access to billing information.[62] National programmes are also to be implemented for large companies to undertake, and to encourage SMEs to undertake, energy audits for identifying ways to improve energy efficiency.[63]

Member states must also establish an EEOS to ensure that energy distributors and/or retail energy sales companies in each member state's territory achieve a cumulative 1.5% end-use energy savings per year through energy efficiency measures.[64] States have discretion to exclude certain sectors, such as transport, which if done has the effect of reducing the annualized energy savings. As of April 2016, only 16 member states had existing or planned EEOS, despite the evidence of their benefit.[65] The European Parliament has found that success factors for the EEOS include usage of market-based solution, credible compliance mechanisms, and a combination of flexible banking and borrowing tools with properly implemented trading mechanisms. The principal factors for failure include unnecessary complexity and lack of political will, government capacity, support instruments for obligated parties, compliance mechanisms and suitable monitoring mechanisms.[66]

In July 2017, the European Commission introduced a new energy labelling regulation for energy efficient products, which will replace its predecessor (European Commission, Energy efficiency products).

New technologies

The use and development of new sustainable energy technologies in Europe, including transportation and battery storage, are governed by several regulatory measures.

In 2009, the EU introduced the Clean Vehicles Directive to 'stimulate the market for clean and energy-efficient road transport vehicles', which will in turn contribute to the EU's environment, climate and energy policies and obligations.[67] It does this by mandating that public authorities and public transport operators procure energy-efficient road transport vehicles that meet technical energy and environmental performance standards.[68] Energy and environmental impacts linked to the operation of vehicles over their whole lifetime, including energy consumption, carbon emissions and pollution, are to be taken into account.[69]

Although the directive does not expressly address electric vehicles, 2008 EU policy, including the European Economic Recovery Plan and the Green Car Initiative, support technological optimization and market development of battery-powered and hybrid vehicles, as well as other green propulsion transports based on biofuels and hydrogen fuel cells (European Commission, Electric vehicles, 2017). Relatedly, the 2009 Fuel Quality Directive (as amended) seeks to regulate petrol, diesel and gas as fuels, including by mandating lower greenhouse gas intensity of energy supplied for road transport, thereby reducing emissions. It also establishes biofuel sustainability criteria for contributions to those intensity reduction obligations.[70]

These regulations, in concert with the regulation of renewable energy, energy efficiency and greenhouse gas emission, favour the development of environmentally friendly transport, including electric vehicles. In 2014, the EU mandated the deployment and installation of alternative fuels infrastructure, including the development of national targets and national policy frameworks for publicly accessible recharging stations or points for electric, liquified natural gas and hydrogen-based vehicles.[71] In 2016, after the Paris Agreement, European cities and regions, transport manufacturers and transport organizations instigated an initiative to increase the deployment of clean energy busses (European Commission, European clean bus deployment initiative, 2017).

Unlike clean transport, there is currently minimal regulation of energy storage (i.e. batteries) in the EU, creating uncertainty for investments. The Clean Energy for All Europeans package proposes changes to the internal market Electricity Directive and the Renewables Directive to better facilitate the use, development and markets for energy storage (European Commission, 30 November, 2016a). Storage technology innovation is also being promoted by various EU programmes (see, e.g., European Commission, Horizon 2020). In February 2017, the European Commission published a Staff Working Document specifically addressing energy storage, which outlines the role of energy storage in relation to electricity, presents the advantages of different technologies and innovative solutions in different contexts, and discusses further possible policy approaches.[72]

Conclusion

Sustainable energy governance is largely decentralized. Traditionally considered a matter of domestic law and policy, the regulation of sustainable energy has escalated to the international arena following growing concerns over serious environmental issues and energy security. It has also escalated at the regional level. As the EU example shows, this allows nations to pursue a path to sustainability based on their own circumstances, while still conforming to a common regulatory framework. The proliferation of international agreements in the last decade is indicative of the fast-developing area, where transparency and technology transfer are key components. Yet international regulation only imposes targets to be achieved in energy-related areas, leaving the policies and measures for reaching such targets to the state parties mindful of their economic development.

The EU has been a leading player, regulating areas spanning renewables to energy efficiency and hybrid vehicles. However, certain areas, such as mass battery storage, fail to be specifically regulated, creating an environment of uncertainty and potentially hindering development in the area, which merits further enquiry. Another area of future research is the role and impact of multi-stakeholder partnerships in influencing global and particularly European sustainable energy transition and related legal frameworks. And while it is beyond the scope of this chapter, consideration of the legal interlinkages between international and regional human rights law and energy access and use is also vital, particularly in the context of the Sustainable Development Goals. Overall, however, the imperative in Europe is not to establish new norms and frameworks, but to continue improving an already exemplary regime.

Notes

1 The views expressed in this chapter, and any errors, are solely those of the author. Special thanks are owed to Marco Paoletti for his helpful research assistance and insightful comments on earlier drafts.

2 Trail Smelter Case (United States of America v Canada) (Awards) (1938/1941) 3 RIAA 1905.

3 *Report of the United Nations Conference on Environment and Development*, UN Doc A/CONF.151/26 (Vol. I) (12 August 1992) annex I [2] ('Rio Declaration'); see generally Viñuales 2015.

4 Principle 8 of the Rio Declaration.

5 *Report of the United Nations Conference on Environment and Development*, UN Doc A/CONF.151/26/Rev.1 (Vol. II) (1993) annex II [9.12(a)], [9.12(d)], [9.12(f)].

6 *Programme for the Further Implementation of Agenda 21*, GA Res S/19–2, UN GAOR, 19th spec sess, 11th plen mtg, Agenda Item 8, Supp No 49, UN Doc A/RES/S-19/2 (19 September 1997) annex [45], [46(a)]–[46(e)], [52].

7 *Report of the World Summit on Sustainable Development*, UN Doc A/CONF.199/20 (2002) ch I pt 2 annex ('Plan of Implmentation').
8 Plan of Implementation, Art. 20(e).
9 International Year of Sustainable Energy for All, UN Doc A/RES/65/151, 2 para 1; *Promotion of New and Renewable Sources of Energy: Report of the Secretary-General*, UN GAOR, 66th sess, Agenda Item 19(j), UN Doc A/66/306 (15 August 2011) 17 [56], 18 [57]–[60], 22 [82].
10 *United Nations Framework Convention on Climate Change*, opened for signature 9 May 1992, 1771 UNTS 107 (entered into force 21 March 1994) Art. 2 ('UNFCCC').
11 UNFCCC, Arts 3, 4(1)(a)–(b).
12 UNFCCC, Art. 4(1)(c).
13 *Kyoto Protocol to the United Nations Framework Convention on Climate Change*, opened for signature 11 December 1997, 2303 UNTS 162 (entered into force 16 February 2005) Art. 2(l)(a)(i), (iv) ('Kyoto Protocol').
14 Kyoto Protocol, Arts 3, 4, 6, 12.
15 See, e.g., Kyoto Protocol, Art. 12.
16 UNFCCC, 'Annual Report of the Executive Board of the Clean Development Mechanism to the Conference of the Parties serving as the Meeting of the Parties to the Kyoto Protocol' (12 November 2015) UN Doc FCCC/KP/CMP/2015/5, para. 14; Kyoto Protocol, Arts 2(1)(a)(iv); Netto and Schmidt 2005, 192.
17 Freestone and Streck (2009); Bali Action Plan, Decision l/CP.13, UN Doc FCCC/CP/2007/6/Add.1 (15 December 2007), para. 2; Conference of the Parties, United Nations Framework Convention on Climate Change, Report of the Conference of the Parties on Its Fifteenth Session, Held in Copenhagen from 7 to 19 December 2009 – Addendum – Part 2: Action Taken by the Conference of the Parties at Its Fifteenth Session, UN Doc FCCC/CP/2009/11/Add.1 (30 March 2010) Decision 2/CP.15 [l]–[2].
18 See, e.g., UNFCCC, Technology mechanism; UNFCCC, Climate finance; CTCN, Climate Technology Centre and Network; UNFCCC Decision 3/CP.17, 'Launching the Green Climate Fund, Annex: Governing instrument for the Green Climate Fund' (15 March 2012) UN Doc FCCC/CP/2011/9/Add.1, [37]–[38].
19 *Paris Agreement*, opened for signature 12 December 2015 (entered into force 4 November 2016) Arts 2(1)(a); 4(1).
20 *Paris Agreement*, Art. 2(1)(c).
21 Paris Agreement, Preamble.
22 *United Nations Convention on the Law of the Sea*, opened for signature 10 December 1982, 1833 UNTS 3 (entered into force 16 November 1994) Arts 56(1)(a), 60, 80 ('UNCLOS').
23 See, e.g., UNCLOS, Arts 2, 18, 19, 56.
24 UNCLOS, Art. 56(1).
25 Convention on Access to Information, Public Participation in Decision-making and Access to Justice in Environmental Matters, opened for signature 25 June 1998, 2161 UNTS 447 (entered into force 30 October 2001).
26 *Convention on Environmental Impact Assessment in a Transboundary Context*, opened for signature 25 February 1991, 1989 UNTS 309 (entered into force 10 September 1997).
27 *Protocol on Strategic Environmental Assessment*, opened for signature 21 May 2003, 2685 UNTS 140 (entered into force 11 July 2010).
28 *Convention on Biological Diversity*, opened for signature 5 June 1992, 1760 UNTS 79 (entered into force 29 December 1993).
29 *Convention on the Conservation of Migratory Species of Wild Animals*, opened for signature 23 June 1979, 1651 UNTS 356 (entered into force 1 November 1983).

30 *Convention on Wetlands of International Importance especially as Waterfowl Habitat*, opened for signature 2 February 1971, 996 UNTS 245 (entered into force 21 December 1975).
31 *Energy Charter Treaty*, opened for signature 17 December 1994, 2080 UNTS 95 (entered into force 16 April 1998) Art. 19(1) ('ECT').
32 ECT, Art. 19(1)(a).
33 ECT, Art. 19(1)(a).
34 *Energy Charter Protocol on Energy Efficiency and Related Environmental Aspects*, opened for signature 17 December 1994, 2081 UNTS 3 (entered into force 16 April 1998) Preamble ('PEEREA').
35 PEEREA, Art. 1.2.a.
36 PEEREA, Art. 1.2.c.
37 PEEREA, Art. 1.2.b.
38 *International Energy Charter*, opened for signature 20 May 2015, Hague II Conference, Objectives ('IEC').
39 IEC, Preamble.
40 IEC, Objectives, para 3.
41 *Statute of the International Renewable Energy Agency*, opened for signature 26 January 2009, 2700 UNTS 45 (entered into force 8 July 2010) Art. II ('IRENA Statute').
42 Art. IV(A)(1), IX(A)(3)(a)–(b), IX(G), X(F), XI IRENA Statute.
43 Report of the Secretary General, 'Promotion of new and renewable sources of energy' A/71/220 (27 July 2016), [78].
44 The Gleneagles Communiqué (July 2005) Climate Change, Energy and Sustainable Development, 3 [11(a)].
45 European Commission (December 2011) 'Energy Roadmap 2050' COM(2011) 8885 final.
46 Directive 2009/28/EC of 23 April 2009 on the promotion of the use of energy from renewable sources and amending and subsequently repealing Directives 2001/77/EC and 2003/30/EC (Renewables Directive) [2009] OJ L140.
47 Renewables Directive, Arts 17, 18.
48 Renewables Directive, Arts 1, 3.
49 Renewables Directive, Arts 2(k), 3(3)(a).
50 Renewables Directive, Art. 14(1)–(2).
51 Renewables Directive, Arts 6, 11.
52 Renewables Directive, Arts 7, 9.
53 Renewables Directive, Arts 4, 15.
54 Renewables Directive, Arts 4(1).
55 Renewables Directive, Art. 16(1).
56 Renewables Directive, Art. 13.
57 Directive 2012/27/EU of 25 October 2012 on energy efficiency, 315/2 Art. 1(1).
58 Energy Efficiency Directive, Art. 3.
59 Energy Efficiency Directive, Art. 14.
60 Energy Efficiency Directive, Art. 15.
61 Energy Efficiency Directive, Arts 4–6; Directive 2010/31/EU of 19 May 2012 on the energy performance of buildings (recast).
62 Energy Efficiency Directive, Arts 9–11.
63 Energy Efficiency Directive, Art. 8.
64 Energy Efficiency Directive, Art. 7.
65 European Parliament (April 2016) 'Implementation of the Energy Efficiency Directive (2012/27/EU): Energy Efficiency Obligation Schemes – European Implementation Assessment', 16.
66 European Parliament (April 2016) 'Implementation of the Energy Efficiency Directive (2012/27/EU): Energy Efficiency Obligation Schemes – European Implementation Assessment', 17.

67 Directive 2009/33/EC of 23 April 2009 on the promotion of clean and energy-efficient road transport vehicles, OJ L 120/5, Preamble [11], [36].
68 Clean Vehicles Directive, Arts 1, 3, 4, 5(3).
69 Clean Vehicles Directive, Art. 5(1)–(2).
70 Directive 98/70/EC, OJ L 140, 5 June 2009, regarding the specification of petrol, diesel and gas-oil and introducing a mechanism to monitor and reduce greenhouse gas emissions.
71 Directive 2014/94/EU of 22 October 2014 on the deployment of alternative fuels and infrastructure OJ L 307/1.
72 European Commission, 'Energy storage – The role of electricity' SWD(2017) 61 final.

References

Boute, A. (2013). "The European Union's shaping of an international law on energy efficiency," in: D. Kochenov and F. Amtenbrink (eds), *The European Union's Shaping of the International Legal Order*. Cambridge University Press, Cambridge, UK, pp. 238–260.

Bradbrook, A.J. (2008). "The development of renewable energy technologies and energy efficiency measures through public international law," in: D. Willman, et al. (eds), *Beyond the Carbon Economy: Energy Law in Transition*. Oxford University Press, Oxford, pp. 109–131.

Bradbrook, A.J. and Wahnshafft, R.D. (2005). "International law and global sustainable energy production and consumption," in: A.J. Bradbrook, et al. (eds), *The Law of Energy for Sustainable Development*. Cambridge University Press, Cambridge, UK, pp. 181–201.

Bruce, S. (2013). "International Law and Renewable Energy: Facilitating Sustainable Energy for All?" *Melbourne Journal of International Law* 14, 1.

Bruce, S. and Stephenson, S. (2016). "SDG 7 on Sustainable Energy for All: Contributions of International Law, Policy and Governance" *UNEP/CISDL Issue Brief 2016*, 1–13 (available at SSRN: https://ssrn.com/abstract=2887694).

CTCN Climate Technology Centre and Network. www.ctc-n.org (accessed 8 November 2017).

Dupuy, P.-M. and Viñuales, J.E. (eds), (2013). *Harnessing Foreign Investment to Promote Environmental Protection: Incentives and Safeguards*. Cambridge University Press, Cambridge, UK.

European Commission, 2030 Climate and Energy Framework. https://ec.europa.eu/clima/policies/strategies/2030_en (accessed 8 November 2017).

European Commission, 2050 Low-Carbon Economy. https://ec.europa.eu/clima/policies/strategies/2050_en (accessed 8 November 2017).

European Commission. Electric Vehicles. https://ec.europa.eu/transport/themes/urban/vehicles/road/electric_en (accessed 8 November 2017).

European Commission. Energy Efficiency Products. https://ec.europa.eu/energy/en/topics/energy-efficiency/energy-efficient-products (accessed 8 November 2017).

European Commission. European Clean Bus Deployment Initiative. https://ec.europa.eu/transport/themes/urban/cleanbus_en (accessed 8 November 2017).

European Commission. Horizon 2020. http://ec.europa.eu/programmes/horizon2020/ (accessed 8 November 2017).

European Commission. (30 November 2016a) Commission Proposes New Rules for Consumer Centred Clean Energy Transition. https://ec.europa.eu/energy/en/news/commission-proposes-new-rules-consumer-centred-clean-energy-transition (accessed 8 November 2017).

European Commission. (30 November 2016b). Putting Energy Efficiency First: Consuming Better, Getting Cleaner. http://europa.eu/rapid/press-release_MEMO-16-3986_en.htm (accessed 8 November 2017).

European Commission. (1 February 2017). Fact Sheet: Renewables: Europe on Track to Reach its 20% Target by 2020. http://europa.eu/rapid/press-release_MEMO-17-163_en.htm (accessed 8 November 2017).

Figueres, C. (2017). "Foreword," in: D. Klein, et al. (eds), *The Paris Agreement on Climate Change: Analysis and Commentary*. Oxford University Press, Oxford, pp. v–vii.

Freestone, D. and Streck, C. (eds), (2009). *Legal Aspects of Carbon Trading: Kyoto, Copenhagen, and Beyond*. Oxford University Press, Oxford.

G7. (2015). Leaders' Declaration. www.g8.utoronto.ca/summit/2015elmau/2015-G7-declaration-en.html (accessed 8 November 2017).

G20. (2017). Hamburg Climate and Energy Action Plan for Growth. www.g20.utoronto.ca/2017/2017-g20-climate-and-energy-en.pdf (accessed 8 November 2017).

Happ, R. (2002). "Dispute Settlement under the Energy Charter Treaty." *German Yearbook of International Law* 45, 331–362.

Hodas, D. (2011). "International law and sustainable energy: a portrait of failure," in: J. Benidickson, et al. (eds), *Environmental Law and Sustainability after Rio*. Edward Elgar, Cheltenham, UK, pp. 257–278.

IEA. (2017). *Energy Efficiency 2017*. OECD/IEA.

Kochenov, D. and Amtenbrink, F. (2014). *The European Union's Shaping of the International Legal Order*. Cambridge University Press, Cambridge, UK.

Leal-Arcas, R. and Minas, S. (2016). "Mapping the International and European Governance of Renewable Energy." *Yearbook of European Law* 35, 1, 1–46.

Najam, A. and Cleveland, C. (2005). "Energy and sustainable development at global environmental summits: an evolving agenda," in: L. Hens and B. Nath (eds), *The World Summit on Sustainable Development*. Springer, The Netherlands, pp. 113–134.

Netto, M. and Schmidt, K.-U. B. (2005). "CDM project cycle and the role of the UNFCCC Secretariat," in: D. Freestone and C. Streck (eds), *Legal Aspects of Implementing the Kyoto Protocol Mechanisms: Making Kyoto Work*. Oxford University Press, Oxford, pp. 177–192.

Oniemola, P.K. (2013). "International Law on Renewable Energy: The Need for a Worldwide Treaty." *German Yearbook of International Law* 56, 239–272.

Redgwell, C. (2006). "International soft law and globalization," in: B. Barton, et al. (eds), *Regulating Energy and Natural Resources*. Oxford University Press, Oxford, pp. 89–107.

Redgwell, C. (2016). "International regulation of energy activities," in: M.M. Roggenkamp, et al. (eds), *Energy Law in Europe*, 3rd edn. Oxford University Press, Oxford, pp. 13–137.

Reed, L. and Martinez, L. (2008). "The Energy Charter Treaty: An Overview." *ILSA Journal of International & Comparative Law* 14, 2, 405–439.

UNEP. (2017). *The Emissions Gap Report 2017: A UN Environment Synthesis Report*. UNEP 23. https://wedocs.unep.org/bitstream/handle/20.500.11822/22070/EGR_2017.pdf?sequence=1&isAllowed=y (accessed 3 November 2017).

UNFCCC. Technology Mechanism. http://unfccc.int/ttclear/support/technology-mechanism.html (accessed 8 November 2017).

UNFCCC. Climate Finance. http://unfccc.int/cooperation_and_support/financial_mechanism/items/2807.php. (accessed 8 November 2017).

Viñuales, J.E. (ed.), (2015). *The Rio Declaration on Environment and Development*. Oxford University Press, Oxford.

7 Towards the East

The Energy Community and the extension of EU climate governance

Stephen Minas[1]

Introduction

As a project in regional integration, the European Union is unique in its reliance on law to achieve harmonization between Member States. Over the years, the Member States and institutions of the EU have created a sophisticated network of laws, at both Union and Member State level, which governs a vast array of activities. This legal enterprise also has an important external dimension. The achievement of a common legal framework governing the world's largest single market gives weight to the EU's legal order in the dealings of EU parties with third parties. This is particularly important in a multipolar international system which is increasingly shaped by continental economies with large populations, and in which '[t]here are two kinds of Member States in Europe, small ones, and those who don't know yet they are small' (Timmermans 2017).

The influence of the EU's legal order in its external dealings has been particularly evident in the case of climate change. To take two examples, the EU Emissions Trading System has been influential in shaping global institutions and markets for carbon pricing and EU standards have acquired extraterritorial legal force through global supply chain contracting. The EU is also extending its climate change legal order to third countries more directly, through treaty, principally through the Agreement on the European Economic Area. This chapter will examine opportunities to utilize treaty law to extend the reach of EU climate governance to the jurisdictions of Eastern Europe and the Caucasus which are members of the Energy Community. The Energy Community has already achieved significant, though uneven, progress in integrating its Contracting Parties into the energy governance of the EU. As this energy governance is increasingly refocused on the transition to a low-carbon economy, the opportunities grow for Energy Community Contracting Parties to converge with the EU on climate governance.

It will be argued that the Energy Community has the potential to be a significant force multiplier of EU climate diplomacy, while simultaneously assisting its Contracting Parties to engage in progressively more ambitious implementation of the Paris Agreement. However, in order for this potential to be realized, the recent promising efforts to expand the Energy Community's renewable energy and climate action functions must be built upon, particularly in the context of the governance of the Energy Union.

The Energy Community and its legal framework

The European Union has elaborated an increasingly ambitious body of laws concerning energy, embodied in Article 194 of the Treaty on the Functioning of the European Union (TFEU) which promotes a 'triple functionality' of internal market liberalization, security of supply and environmental sustainability (Buschle 2011, p. 303). Climate technology in the guises of renewable energy, energy efficiency and digitalization is critical to the attainment of environmentally sustainable energy in accordance with the TFEU and the broader energy *acquis communautaire*. To that end, the EU has long negotiated under the United Nations Framework Convention on Climate Change (UNFCCC) in favour of ambitious climate targets and frameworks that are consistent with the EU's internal transition to environmentally sustainable energy and industry.

In addition to this overarching negotiating stance in multilateral climate *diplomacy*, the EU has also worked to directly extend its frameworks concerning climate change or sustainable energy to non-EU Member States through *law*. This has been progressed through other treaty processes, namely the Agreement on the European Economic Area (EEA) and the Treaty establishing the Energy Community, of which this chapter is concerned with the latter. These legal mechanisms complement EU diplomacy under the UNFCCC by extending the operation of relevant EU laws, and the norms which they embody, to third countries, thereby expanding the community of interest with a stake in the EU's broad approach to climate change mitigation, while also potentially converging the UNFCCC negotiating positions of third countries with that of the EU.

The Energy Community was established by Treaty concluded in 2005 (entering into force in 2006) with the objective of integrating the non-EU Member States of South East Europe into the EU's internal energy market (Karova 2015, 23). The Energy Community Treaty tasks the Community with establishing a 'single regulatory space' for the trade in network energy (Article (1)(b)), encompassing the EU and Contracting Parties. The mechanism of achieving integration is the transposition of EU energy legislation into the laws of Energy Community Contracting Parties, generally with extended deadlines and modifications to accommodate national circumstances. Subsequently, the Energy Community expanded beyond South East Europe to include European Neighbourhood Policy countries such as Ukraine and Moldova, consistent with the Council of the EU's statement that the 'Energy Community should be promoted as a framework for energy relationships with countries in the Western Balkans, Eastern Europe and other neighbouring countries willing and able to implement the relevant EU acquis' (2013, p. 8).

Today, the European Union, represented by the Commission, is a Party to the Energy Community and a permanent Vice-President. The nine Contracting Parties to the Energy Community are not EU Member States, although some are candidate countries for accession to the EU. The Contracting Parties are Albania, Bosnia and Herzegovina, Kosovo,[2] the former Yugoslav Republic of Macedonia, Georgia, Moldova, Montenegro, Serbia and Ukraine. Georgia is the most recent Contracting Party, having joined in July 2017. In addition, Armenia, Norway and Turkey are Observers, while any EU Member State has the right to be a Participant in Energy Community meetings. Twenty EU Member States have exercised this right.

The designated activities of the Energy Community are the 'implementation by the Contracting Parties of the *acquis communautaire* on energy, environment, competition and renewables, as described in Title II of the Energy Community Treaty, adapted to both the institutional framework of the Energy Community and the specific situation of each of the Contracting Parties', the establishment of a 'specific regulatory framework' for network energy market operation across the Contracting Parties and the EU, and the creation of 'a market for Network Energy without internal frontiers' for the Parties (Energy Community Treaty, Article 3).

The Energy Community's institutions are obliged to 'interpret any term or other concept used in this Treaty that is derived from European Community law in conformity with the case law of the Court of Justice or the Court of First Instance of the European Communities', where such an interpretation is available (Energy Community Treaty, Article 94). The Energy Community is, therefore, 'an instrument of EU external energy relations' (Talus 2016, p. 261). However, it is also a 'win-win instrument' (High Level Reflection Group 2014, p. 8), contributing to the enabling environment for investment and infrastructure development in Contracting Parties even as it bolsters EU security of energy supply and extends the operation of the EU legal order.

Energy Community governance is channelled through several key bodies. The Ministerial Council, consisting of one representative of each Contracting Party and two representatives of the European Community, is responsible for ensuring the achievement of Energy Community objectives, including through providing general policy guidelines, taking Measures and adopting Procedural Acts (Energy Community Treaty, Articles 47–48). The Permanent High Level Group is responsible for preparing Ministerial Council meetings and reporting to the Ministerial Council on progress, *inter alia* (Article 53). The Secretariat supports the work of the Ministerial Council, Permanent High Level Group and other Community bodies and reviews Party implementation of Treaty obligations, *inter alia* (Article 67). To implement the energy *acquis* in Contracting Parties, the Ministerial Council, Permanent High Level Group or Regulatory Board take Measures following a proposal by the European Commission (Article 79). Measures can be in the form of either 'legally binding' Decisions or non-binding Recommendations (Article 76).

Energy Community law relevant to climate technology

Environmental objectives are prominent in the Energy Community's legal framework. The Energy Community Treaty tasks the Community with 'improv-[ing] the environmental situation with regard to Network Energy and related energy efficiency, fostering the use of renewable energy, and setting out the conditions for energy trade in the single regulatory space' (Article 2(1)(d)). Chapter III of Title II of the Treaty details the environmental *acquis* that Contracting Parties are obliged to implement. This does not include measures specifically targeting climate change.

The energy *acquis* which Contracting Parties are obliged to implement is specified in Annex I of the Treaty and includes EU Directives on the internal markets in electricity and in gas, on conditions for access to electricity and gas networks, and on measures concerning safeguarding of electricity and gas supply. The energy efficiency directives (2012/27/EU, 2010/30/EU and 2010/31/EU) are included in the Energy Community's *acquis*.

In addition, Contracting Parties are obliged to implement Directive 2009/28/EC on the promotion of the use of energy from renewable sources (Renewable Energy Directive or RED) (Energy Community Treaty, Article 20), following the Ministerial Council's 2012 Decision to amend the Treaty in response to the creation of the RED, which replaced previous Directives relating to renewables (Ministerial Council 2012).

The Energy Community Treaty further provides that the Community 'may adopt Measures to foster development in the areas of renewable energy sources and energy efficiency, taking account of their advantages for security of supply, environmental protection, social cohesion and regional development' (Article 35). The Treaty also empowers the Ministerial Council, by unanimity, to 'decide to implement other parts of the *acquis communautaire* related to Network Energy' (i.e. electricity, gas and oil) or to 'extend this Treaty to other energy products and carriers or other essential network infrastructures' (Article 100).

There is currently no binding Energy Community law explicitly focused on climate change (Feiler and Vajda 2017, p. 442). The absence of legal obligations on Contracting Parties concerning climate change, despite the fact that the emissions-intensive electricity, oil and gas sectors are the core business of the Energy Community, is an obvious 'gap' in its legal framework (Energy Community Secretariat 2016b). Article 13 of the Energy Community Treaty provides that '[t]he Parties recognise the importance of the Kyoto Protocol. Each Contracting Party shall endeavour to accede'. This is the only reference to UNFCCC law in the Treaty, which requires updating to reflect that (1) Kyoto's second commitment period (which has not acquired legal force due to lack of ratifications) is due to end in 2020; (2) the Paris Agreement, in force since 2016, has become the primary multilateral mechanism for implementing the UNFCCC objectives. Indeed, members of the Energy Community Secretariat have warned that if this provision is not updated, there would be 'serious risk for the Energy Community to be decoupled from the global commitment to fight climate change' (Buschle and Lazzerini 2017, p. 33).

The development of climate-relevant work under the Energy Community

Despite the as-yet-incomplete incorporation of climate-relevant EU law into Energy Community law, recent years have seen a significant expansion of Energy Community activities aimed at improving the environmental sustainability of energy in Contracting Parties. The Secretariat both works with Contracting Parties to facilitate domestic implementation of EnC law and takes action to enforce compliance with EnC law within the Contracting Parties. Sustainable energy is increasingly featuring in both facilitation and compliance activities.

The High Level Reflection Group (HLRG), convened by the Ministerial Council in 2013 to assess the effectiveness of the Community, identified a 'lack of ambition in the areas of environmental protection and climate change' (High Level Reflection Group 2014, p. 12). To remedy this, the Group proposed consideration of incorporating additional EU environmental directives into Energy Community law, including Directive 2003/87/EC establishing the EU Emissions Trading System (ETS). The extension of emissions trading to the Contracting Parties would pose significant legal and administrative challenges for the Energy Community (Vajda 2015), and the proposal has not been taken up yet. Nevertheless, the HLRG's identification of the need for the Community to do more in the climate area, when coupled with the subsequent Paris Agreement and the EU's adoption of its 2030 targets and Energy Union package, succeeded in placing enhanced climate action on the agenda.

Much of the progress in initiating Energy Community activity in respect of climate change has pertained to only a subset of the Contracting Parties, being the six jurisdictions of the Western Balkans. Through the Western Balkan 6 Initiative (WB6), launched in 2014, the Secretariat works with the Western Balkans Contracting Parties to strengthen energy cooperation in a range of areas, including sustainability. In July 2016, the WB6 jurisdictions adopted a Sustainability Charter, in which they each commit to a variety of actions to enhance frameworks to promote energy efficiency, develop support measures to improve the sustainability of their energy systems (such as market mechanisms) and implement 'robust' greenhouse gas emissions monitoring and reporting (WB6/Energy Community 2016). Each commitment has a corresponding deadline. The WB6 countries asked the Secretariat to assist them in implementing the Charter and to monitor progress. One of the stated purposes of the Charter is to assist WB6 countries to meet their commitments under the Paris Agreement. In its November 2016 communication 'Clean Energy For All Europeans', the European Commission committed to further strengthen energy cooperation with the Western Balkans (European Commission 2016b, p. 13).

In its most recent Sustainability Charter monitoring report of June 2017, the Secretariat found variable progress. All countries had advanced governance frameworks for energy efficiency, e.g. through transposition of the Energy Performance of Buildings Directive. Progress on support measures for renewable energy deployment was reported in some countries, while all jurisdictions supported incorporated a mechanism for monitoring and reporting emissions into Energy Community law and were 'generally' fulfilling their UNFCCC reporting obligations. However, investment promotion was unsatisfactory and potential investors in sustainable energy continued to confront 'highly bureaucratic administrative systems' (Energy Community Secretariat 2017a).

The Energy Community's dispute resolution function has also contributed to the governance of sustainable energy in Contracting Parties. Notwithstanding the continuing absence of a judicial dispute resolution mechanism (see Title VII of the Energy Community Treaty and Buschle 2016), the Secretariat has used existing dispute settlement procedures to promote Contracting Party compliance with renewable energy obligations. In October 2014, the Secretariat initiated proceedings against Ukraine over Ukraine's Electricity Law, which made eligibility for feed-in tariffs dependent on minimum levels of local content. The Secretariat's Opening Letter alleged that this local content requirement was in breach of both the Renewable Energy Directive and the Energy Community Treaty's Article 7 prohibition on discrimination. In response, Ukraine's Parliament removed the local content threshold for FiT eligibility, but made investors meeting local content quota eligible for an additional 'bonus' payment, thereby 'replacing a strict barrier to market access by a more differentiated instrument' (Energy Community, Case ECS 07/13). As a result, the Secretariat closed the case.

More recent developments in the Energy Community's dispute resolution architecture have also been applied to renewable energy matters. In 2016, the Secretariat established a new Dispute Resolution and Negotiation Centre in order to provide for 'rational, neutral, quick, efficient and sustainable resolution' of energy disputes (Energy Community Secretariat 2016c). The Centre offers facilitated negotiation and mediation of disputes, including those concerning states, national authorities and private parties. In one of the earliest matters dealt with, the Centre mediated a dispute between the Albanian government and renewable energy producers, resulting in Albania adopting a Law on Promotion of the Use of Energy from Renewable Sources, 'including a clause on the compensation of existing renewables producers' (Energy Community Secretariat 2017a).

Despite these positive developments, there remains a significant focus within most Contracting Parties on fossil fuel capacity. Moreover, the relatively high renewable energy potential of Contracting Parties is offset by a higher cost of capital than within the EU (Feiler and Vajda 2017, pp. 442–443). It is clear that substantial reform is necessary both at the levels of the Energy Community and individual Contracting Parties to truly place these countries onto the path of sustainable energy development.

Post-Paris and towards 2030: activities to strengthen climate action under the Energy Community

Since the adoption of the Paris Agreement in 2015, there has been new momentum within Energy Community processes to strengthen legal frameworks and implementation concerning climate change. This momentum has encompassed proposals to amend the Energy Community Treaty to better reflect the need for Contracting Parties to take action on climate change, moves to incorporate emissions monitoring into Energy Community law and the establishment of the Climate Action Group to institutionalize cooperation in this area.

These moves toward harmonization and cooperation between the EU and Contracting Parties on climate change can take advantage of a new equivalence in EU and Contracting Parties obligations under the Paris Agreement. Until the 2016 entry into force of the Paris Agreement, the Contracting Parties and the EU under international climate law were subject to different treatment under international climate law. Most Contracting Parties generally had less obligations under the UNFCCC and its Kyoto Protocol than did the countries that are now Member States of the EU. Of the current Contracting Parties, only Ukraine is included in the Annex I list of UNFCCC Parties with additional obligations (and, even so, Ukraine's status as a post-Communist economy-in-transition without Annex II obligations is acknowledged in the Climate Convention), while other Contracting Parties are non-Annex I Parties for the purposes of the Climate Convention. The distinction of annexes is no longer maintained in the Paris Agreement, with the result that Contracting Parties and EU Member States now face the same basic requirement, that of developing, reporting and implementing Nationally Determined Contributions. The adoption by Contracting Parties of EU climate frameworks, therefore, becomes more practicable.

To better reflect this development in international climate law, the Secretariat has proposed revision of Article 13 of The Energy Community Treaty concerning climate change, which 'never gained practical relevance', to read: 'The Parties recognize the importance of the Kyoto Protocol and the Paris Agreement. The Energy Community shall implement them in line with the European Union's policy and legislation' (Energy Community Secretariat 2016a).

More specifically, in October 2016, the Ministerial Council decided to begin a process of incorporating EU Regulation No 525/2013 on a mechanism for monitoring and reporting (MMR) greenhouse gas emissions into Energy Community law (Ministerial Council 2016). Noting that the Regulation would need to be amended for 'full and legally binding incorporation' into the Energy Community, the Ministerial Council requested Contracting Parties to begin legal and institutional preparations and the Secretariat to assist them (Article 1). A final decision on incorporating the Regulation will be subject to a proposal from the European Commission (Article 3). In its 2017 Annual Implementation Report, the Secretariat monitored Contracting Party compliance with Regulation No 525/2013 for the first time and found, in most case, that significant additional work is required to achieve compliance (2017c).

Even in advance of the transposition of new legal responsibilities, the Energy Community is deepening practical cooperation. In June 2017, the Energy Community held its first informal ministerial meeting attended by both the environment and energy ministers of Contracting Parties. This meeting endorsed the creation of a Climate Action Group (CAG) within the Energy Community to 'facilitate the development of climate policy and the transposition of related legislation into national legislation and support its effective implementation and to design 2030 targets for the Energy Community Contracting Parties' (Energy Community Secretariat 2017b). The CAG has attracted high-level participation, being co-chaired by former EU climate commissioner and UNFCCC COP15 president Connie Hedegaard.

The first meeting of the CAG took place in September 2017 with the participation of all Contracting Parties. The Group elected the co-chairs and adopted its rules of procedure and work programme. In addition to discussing climate and energy planning and coordination of negotiating positions under the UNFCCC (both of which are addressed below), the CAG hosted a discussion of 'lessons learnt' in the transition to sustainable energy, with both the EU and Contracting Parties sharing experiences. The CAG aims to facilitate the establishment of 'well-functioning and inclusive processes' for managing the energy transition in each Contracting Party (Energy Community Climate Action Group 2017, p. 2). A further topic addressed by the CAG was the improvement of internal cooperation and capacity-building. This agenda item focused on identifying 'bottlenecks' in policy processes at the national level, particularly concerning coordination between the ministries, respectively, responsible for energy and climate policy. The CAG is identified as 'the driving force of enhancing cooperation between the ministries' (Energy Community Climate Action Group 2017, pp. 3–4). Options canvassed to address the coordination problem included inter-ministerial working groups and memoranda of understanding between relevant ministries. The CAG also considered activity at the Energy Community level to improve coordination, including Secretariat assistance and reporting and joint meetings of the Energy Community's energy efficiency, renewable energy and environmental bodies. Following December 2017's Ministerial Council meeting, the CAG was renamed the Energy and Climate Committee (Ministerial Council 2018, Article 2).

The Energy and Climate Committee can be a particularly important resource for the implementation of the Paris Agreement in Contracting Parties, including through the process of developing progressively more ambitious Nationally Determined Contributions. The success of the Energy and Climate Committee will depend on its ability to effectively address the particular challenges faced by Contracting Parties in developing and implementing climate regulation, including data limitations and regulatory capacity. As identified, these challenges 'should be openly and seriously addressed rather than dodged' (Energy Community Climate Action Group 2017, p. 2).

Technological innovation and the Energy Union

Recent and ongoing innovations in the areas of electricity generation, transmission and storage, and the digitalization of consumption management, have the potential to hasten the global transition to low-emissions energy generation.

These innovations include the development of battery storage capacity, including within electric cars; advances in the interconnection of electricity systems, as through transmission and distribution cabling, that would improve the ability to harness intermittent renewable energy sources; the growing use of broadband and data to actively manage electricity demand and consumption, as through 'smart metres'; and the capacity for distribution generation from renewable energy sources, including by household 'prosumers' (Helm 2017, pp. 62–88; Glachant, Rious and Vasconcelos 2015).

These developments in renewable energy deployment and digitalization within the electricity sector have necessitated a regulatory response to both challenges and opportunities that arise. In the EU, the framework of the Energy Union is the identified mechanism to both manage these technological transitions and harness them for the achievement of the EU's 2030 climate targets. The incorporation of the Contracting Parties of the Energy Community into the still-developing governance framework of the Energy Union constitutes a new opportunity to deepen integration along a sustainable energy trajectory.

In February 2015, following a number of policy proposals from national and institutional actors concerning both energy security and climate policy (Szulecki et al. 2016), the European Commission released its Energy Union Package. The Energy Union responds to the problems of inconsistent national regulatory frameworks, lack of consumer choice and the challenges of modernizing energy infrastructure and leading in low-emissions technology innovation. It sets out to achieve progress in the five dimensions of energy security, market integration, energy efficiency, decarbonization and research, innovation and competitiveness (European Commission 2015, pp. 3–4).

While the Energy Union policy displays significant continuity with previous EU energy and climate policies, it introduces 'changes by layering' new policy components onto familiar material (Pellerin-Carlin 2017, p. 71). Democratization – giving the citizen additional choice in the market, the capacity to manage demand and the ability to produce as well as consume energy – is one such new element. Promotion of citizen empowerment, harnessing improving technologies of distributed generation and smart metering, is included in legislative proposals to implement the Energy Union. For example, the Commission's proposal for a revised Renewable Energy Directive includes new provisions enabling 'renewable self-consumers' to sell excess generated electricity, either individually or through aggregators, while maintaining their consumer rights (Article 21) and providing a specific regime for small-to-medium enterprises and not-for-profit organizations that can be classified as 'renewable energy communities' (Article 22).

The Energy Union Framework calls for an 'integrated governance and monitoring process' which would, *inter alia*, strengthen coherence by bringing together climate and energy actions (European Commission 2015, p. 17). In November 2016, the European Commission released its proposal for the governance of the Energy Union (European Commission 2016a), as part of the 'Clean Energy For All Europeans' package (European Commission 2016b). 'Integrated national energy and climate plans' covering the five dimensions of the Energy

Union are at the centre of the proposed governance arrangements. These plans are to be submitted for 10-year periods beginning with 2021–2030 and would be subject to monitoring by the Commission, with submission of progress reports (Article 1 of the proposed Regulation). The first 10-year period is particularly geared to meeting the EU's 2030 climate targets. Detailed specifications of what the integrated plans are to include are listed in Chapter 2 of the proposed Regulation. Article 11 further specifies that 'Member States shall cooperate with each other at regional level to effectively meet the targets, objectives and contributions set out in their integrated national energy and climate plan'. The proposed Regulation calls on the Commission to assess, every 2 years, both overall progress on meeting the Energy Union objectives and individual Member State progress measured against its integrated plan (Article 25).

From the initiation of the Energy Union process, the involvement of the Energy Community has been contemplated. In the February 2015 Energy Union Package, the European Commission set the target of 'strengthen[ing] the Energy Community, ensuring effective implementation of the EU's energy, environment and competition *acquis*, energy market reforms and incentivizing investments in the energy sector. The goal is closer integration of the EU and Energy Community energy markets' (European Commission 2015, p. 7).

However, the details of how Energy Union governance would be applied to Contracting Parties remain unclear. First, the Energy Union strategy and related 2030 targets are not part of the Energy Community's *acquis* (and, indeed, have yet to be introduced into EU law). Second, even assuming the adoption by the EU of legislation giving effect to the Energy Union and subsequent inclusion into Energy Community law, Contracting Parties face additional challenges, such as the quality of data and interagency cooperation that would be necessary to develop integrated national energy and climate plans.

The recently established Energy and Climate Committee has the potential to facilitate governance in Contracting Parties that converges with the Energy Union framework. At its first meeting, it was agreed that the CAG (as the Energy and Climate Committee then was) would be the 'primary platform' for integrated climate and energy planning in the Energy Community, with each Contracting Party invited to appoint focal points from both ministries responsible for energy and climate policy (Climate Action Group 2017, p. 2). In December 2017, the Ministerial Council adopted a Recommendation inviting Contracting Parties to begin preparations, with the Secretariat's assistance, to adopt integrated national energy and climate plans, addressing each dimension of the Energy Union, for the period 2021-2030 (Ministerial Council 2018, Article 1). Within the Energy and Climate Committee, the Secretariat and European Commission are empowered to identify elements and procedures for both the national plans and progress reports (Article 2).

A further topic that the Energy and Climate Committee could usefully address is the contribution of regional cooperation, both between neighbouring Contracting Parties and as between Contracting Parties and neighbouring Member States, to the achievement of 2030 targets. This is particularly important in the context of EU 2030 targets that are set at the Union level but not specified for individual Member States, contrary to the 2020 targets. In December 2017, the Ministerial Council decided that integrated national plans should promote regional cooperation and should be coordinated to avoid 'perverse incentives' (Article 3).

If effectively implemented, the Energy Union will itself constitute a major boost to EU climate diplomacy, as the 'major vector' of the EU's contribution to the global energy transition (European Commission 2016b, p. 3). The additional credibility that will result from the implementation of the 2030 targets through a legal framework of unprecedented ambition would add ballast to the EU's leading role in climate diplomacy, especially when contrasted to the unravelling by the Trump Administration of Obama-era executive climate actions in the United States. Furthermore, driving convergence of the Contracting Parties with the EU's approach to climate action through Energy Union governance would further demonstrate the EU's ability to lead by example.

The Energy Community and EU climate diplomacy through the UNFCCC

Finally, there is an opportunity for the Energy Community to more directly support EU climate diplomacy through the UNFCCC negotiations, which are the focus of many of the chapters of this book. At its first meeting in September 2017, the Climate Action Group agreed to support information exchange and coordination among Contracting Parties, in order to 'align positions with the EU where possible in order to make the process smoother and having a higher negotiation impact' (Climate Action Group 2017, p. 3). This activity began with the 23rd meeting of the UNFCCC Conference of Parties in Bonn in November 2017, with the organization of pre-COP negotiation training for Contracting Party delegations and a second meeting of the CAG between Contracting Parties and the EU's Directorate-General for Climate Action at the COP, in order to exchange views and coordinate on a number of topics of the Paris work programme.

Already, it has been commonplace for individual Contracting Parties to associate themselves with certain EU positions in the UNFCCC negotiations (for example, a 2017 EU submission concerning the Technology Framework under the Paris Agreement was supported by four Contracting Parties). Full convergence between the negotiating positions of all Contracting Parties and the EU is not, however, currently a realistic possibility. There are, for example, long-held divergences between the EU and the so-called economies in transition such as Ukraine on issues such as response measures and the retiring of emissions allowances. Nevertheless, while it will be challenging for Contracting Parties as a group to coordinate their positions with the EU on all issues within the sprawling UNFCCC negotiation, the goal of 'speaking with one voice' (Climate Action Group, 2017) wherever feasible would both enhance the weight of the EU position and assist Contracting Parties (especially those who have opened EU accession proceedings) to harmonize climate policy with that of the EU.

Conclusion

Aligning the 'legal DNA' of the Energy Community Contracting Parties with that of the EU has always been an ambitious undertaking (Energy Community Secretariat 2017c, p. 8). The increasing urgency of the climate imperative has added complexity to this already difficult endeavour, but also has the potential to create new common ground and opportunities for convergence. The Energy Community is significantly increasing its focus on climate change and sustainable energy. With the stated ambition of developing and implementing climate policy 'based on EU best practice' (Energy Community Climate Action Group 2017, p. 1), the Energy Community illustrates EU influence in climate diplomacy as an acknowledged leader in the field. If the Energy Community can implement a broadened climate *acquis* in its Contracting Parties, including the legal outcomes of the Energy Union policy, while also achieving convergence with EU negotiating positions in the UNFCCC, it will itself become a significant contributor to the reach of EU climate diplomacy.

To maximize these efforts, it will be necessary to overcome the 'growing reluctance' identified by the Secretariat to the proposing of EU energy legislation for adoption into Energy Community law, a reluctance which results in a 'widening of the legal gap' between the EU and Contracting Parties in the context of significant EU legislative activity (Energy Community Secretariat 2017c, p. 8). Two key opportunities have presented themselves. One is to fully implement Energy Union governance into Energy Community law, allowing integrated planning and monitoring of climate change, renewable energy and energy efficiency actions and ending the compartmentalized approach. The other is to amend the Energy Community Treaty to make it fit for purpose for a time when climate change is the central challenge of energy policy, but also when emerging low-emission technologies offer solutions to the 'energy trilemma' of affordable energy in functioning markets, energy security and environmental sustainability.

Notes

1 I am grateful to Irina Lazzerini and Peter Vajda for their helpful comments. All opinions and any errors are my own.
2 This designation is without prejudice to positions on status, and is in line with UNSCR 1244 and the ICJ Opinion on the Kosovo declaration of independence.

References

Buschle, D. (2011). "The enforcement of energy law in wider Europe," in: D. Buschle, S. Hirsbrunner and C. Kaddous (eds), *European Energy Law, Droit Européen de l'Énergie*. Helbing Lichtenhahn, Basel, Switzerland, pp. 301–342.

Buschle, D. (2016). "The Enforcement of European Energy Law Outside the European Union – Does the Energy Community Live Up to the Expectations?" *European Energy Journal* 6, 1, 26–44.

Buschle, D. and Lazzerini, I. (2017). "Energy Community. Driving Sustainability in EU Neighboring Countries." *SETIS Magazine* 15, 32–34.

Council of the European Union, Follow up to the European Council of 22 May 2013; Review of Developments on the External Dimension of the EU Energy Policy, 17756/13, 12 December 2013.

Energy Community, Case ECS 07/13: Ukraine/Renewable Energy. https://energy-community.org/legal/cases/2013/case0713UA.html.

Energy Community Climate Action Group. (2017) 1st Meeting, Vienna, Austria, 5 September 2017, Report.

Energy Community Secretariat. (2016a). Annex 01/41st PHLG/04-03-2016: Proposed Treaty Changes For the Ministerial Council in October 2016.

Energy Community Secretariat. (2016b). Annex 07a/4st PHLG/04-03-2016: Explanatory Note: Proposal for a Decision on the implementation of Regulation (EU) No 525/2013 on a mechanism for monitoring and reporting greenhouse gas emissions and for reporting other information at national and Union level relevant to climate change and amending Article 16 and Annex II of the Energy Community Treaty.

Energy Community Secretariat. (2016c). Procedural Act of the Energy Community Secretariat, 2016/3/ECS on the Establishment of a Dispute Resolution and Negotiation Centre.

Energy Community Secretariat. (2017a). "WB6 Sustainability Charter Monitoring Report."

Energy Community Secretariat. (2017b). Energy Community Climate Action Group Agrees on Pathway to Meet Climate Change Challenges. www.energy-community.org/news/Energy-Community-News/2017/09/05.html.

Energy Community Secretariat. (2017c). Annual Implementation Report.

European Commission. (2015). "Energy Union Package: A Framework Strategy for a Resilient Energy Union with a Forward-Looking Climate Change Policy," COM(2015) 80 final.

European Commission. (2016a). "Proposal for a Regulation of the European Parliament and of the Council on the Governance of the Energy Union," COM(2016) 759 final.

European Commission. (2016b). "Clean Energy For All Europeans," Communication from the Commission to the European Parliament, the Council, the European Economic and Social Committee, the Committee of the Regions and the European Investment Bank, COM(2016) 860 final.

Feiler, J. and Vajda, P. (2017). "Energy and environment," in: R. Leal-Arcas and J. Wouters (eds), *Research Handbook on EU Energy Law and Policy*. Edward Elgar, London, pp. 432–449.

Glachant, J.M., Rious, V. and Vasconcelos, J. (2015). "What future(s) for the EU power transmission industry?" Florence School of Regulation Policy Brief 2015/04.

Helm, D. (2017). *Burn Out: The Endgame for Fossil Fuels*. Yale University Press, New Haven, CT.

High Level Reflection Group of the Energy Community. (2014). "An Energy Community for the Future."

Karova, R. (2015). "The dispute settlement system of the energy community: testing its effectiveness," in: D. Buschle and K. Talus (eds), *The Energy Community: A New Energy Governance System*. Intersentia, Cambridge, UK, pp. 23–65.

Ministerial Council of the Energy Community. (2016a). D/2012/04/MC-EnC: Decision on the implementation of Directive 2009/28/EC and amending Article 20 of the Energy Community Treaty.

Ministerial Council of the Energy Community. (2016b). Recommendation R/2016/02/MC-EnC on preparing for the implementation of Regulation (EU) No 525/2013 on a mechanism for monitoring and reporting greenhouse gas emissions.

Ministerial Council of the Energy Community (2018). Recommendation 2018/1/MC-EnC on preparing for the development of integrated national energy and climate plans by the Contracting Parties of the Energy Community.

Pellerin-Carlin, T. (2017). "The European Energy Union," in: R. Leal-Arcas and J. Wouters (eds), *Research Handbook on EU Energy Law and Policy*. Edward Elgar, London, pp. 67–102.

Szulecki, K., et al. (2016). "Shaping the 'Energy Union': Between National Positions and Governance Innovation in EU Energy and Climate Policy." *Climate Policy* 16, 5, 548–567.

Talus, K. (2016). "Float Like a Butterfly, Sting Like a Bee: Judicial Challenges to Renewable-Energy Support Schemes in Europe." *Climate Law* 6, 250–263.

Timmersmans, F. (2017). Speech of First Vice-President Frans Timmermans, Future Force Conference. https://ec.europa.eu/commission/commissioners/2014–2019/timmermans/announcements/speech-first-vice-president-frans-timmermans-future-force-conference_en

Treaty Establishing the Energy Community. (2005).

Vajda, P. (2015). "The Environmental Dimension of the Proposals of the Energy Community's High Level Reflection Group." *Oil, Gas and Energy Law* 13, 6, 19–20.

WB6/Energy Community. (2016). "Western Balkan Sustainability Charter."

8 The Investment Plan for Europe

Private capital and climate governance

Megan Bowman[1]

Introduction

Complex and contested legal and governance approaches are emerging in response to the transnational and multi-sectoral impacts of climate change. At the international level, the Paris Agreement to the United Nations Framework Convention on Climate Change (UNFCCC), which came into force in November 2016, acknowledges the global effort required to address climate change. It urges ratifying nations to design their regulatory systems in a way that helps meet the global target of a 'well below' 2°C world (art. 2(1)(a)) through their own Nationally Determined Contributions (NDCs) (art. 3; also arts 4, 7, 9, 10, 11, 13). Moreover, it urges developed countries to 'take the lead in mobilizing climate finance from a wide variety of sources, instruments and channels' and in ways that

- give attention to 'the needs and priorities of developing country Parties' (art. 9(3); also arts 4(5), 9, 10, 11);
- facilitate not only mitigation through lower greenhouse gas (GHG) emissions but also adaptation through 'climate-resilient development' (art. 2(1)(c); also arts 9(1), 9(4), 7); and
- exceed previous efforts at financial mobilization (art. 9(3)).

The combination of these elements is critical: the Paris outcome urges developed countries to mobilize US$100 billion per year by 2020 for climate action in developing countries; yet the actual investment required for a global transition has been estimated at US$90 trillion total, or $5.6 trillion per year, by 2030 (International Energy Agency 2014). Bridging this sizeable gap necessarily requires increased flows of private capital for climate-related purposes. Certainly, there is much hope that the Green Climate Fund, which is an operating entity of the UNFCCC Financial Mechanism alongside the Global Environment

Facility, will support the efforts of developing countries by channelling a share of new multilateral funding, comprising public funds and also private capital through its Private Sector Facility. Yet many are also placing a clear emphasis on non-state actors, such as corporations and private finance actors, to scale up their own efforts and to make decisions and support actions that reduce greenhouse gas emissions and enhance climate resilience (e.g. United Nations 2015; Bowman 2015). Indeed, the *Marrakech Partnership for Global Climate Action* explicitly acknowledges the need for greater public–private cooperation whereby '[private] entities have a key role to play in assisting governments to translate NDCs into investment-ready vehicles as well as to scale up investment in [climate-friendly] infrastructure' (COP22, final (2016)).

The European Union (EU) has a vital role in the climate endeavour not just as an amalgam of the individual efforts of nation states but also as a coherent region. The goals of the EU 2030 climate and energy framework form the basis of the EU NDC under the Paris Agreement (European Parliament 2015; Latvia and European Commission 2015). These goals require that, compared to 1990 levels, the EU as a block ought to reduce greenhouse gas emissions by 40%, increase renewable fuel sources by 27% and increase energy efficiency by 27% (European Commission 2017c). Importantly, following the putative withdrawal by the US from the Paris Agreement, the EU claims to want to take leadership on implementing the global transition (Thimann 2017). This is reinforced by the mandate of the EU High-Level Expert Group on Sustainable Finance (HLEG), which was formed in early 2017 and delivered its interim recommendations in July 2017 with its final report published in January 2018 (European Commission 2016a; High-Level Expert Group on Sustainable Finance 2018). Specifically, HLEG Chair Christian Thimann has described a twofold imperative of 'sustainable finance' as being 'inclusive growth [regarding] long-term needs for innovation and infrastructure, and accelerating the shift to a resource-efficient and low-carbon economy' and also 'a financial system that is focused on the longer term as well as material ESG [environmental, social, governance] factors'. (Thimann 2017, pp. 2–3).

There is clear importance for the European Commission to help *all* EU member states to achieve the objectives of the Paris Agreement and to facilitate the goals of sustainable finance. The key question is how best to do it.

This chapter contends that the Investment Plan for Europe presents an untapped opportunity for channelling a cohesive approach to mobilizing private capital for climate-related infrastructure and investment in Europe and abroad. Specifically, this could occur through the European Fund for Strategic Investment (EFSI), widely known as the 'Juncker Fund', together with the Plan's enabling platforms being the European Investment Project Portal and the European Investment Advisory Hub (EIAH). The Investment Plan for Europe was announced by the European Commission (EC) on 26 November, 2014 (COM(2014) 903 final, 5) as a response to the downturn in intra-European investment following the 2008 global financial crisis. It is not focused on climate-related investment *per se* but rather on enhancing regional development and investment levels in Europe (European Investment Bank n.d.). This chapter contends that the EFSI investment and EIAH informational pillars of the Plan could be improved to facilitate public–private climate-related investment that enhances both regional implementation of, and international leadership in achieving, Paris Agreement goals and, in so doing, potentially enhance sustainable finance.

The remainder of this chapter is structured as follows. Part 2 explains that engaging the private sector for climate-related purposes requires attention to business case logic, and that law and policy (regulation) has a crucial role in shaping corporate and investor cost/benefit analyses. Specifically, it reveals that public–private climate finance regulation comprises two types: first, financial mechanisms, and second, facilitative modalities. This knowledge is then applied to the EFSI in Part 3, which details its purpose and progress before critiquing current decision-making under three core issues of hierarchy, additionality and geographical coverage. Recommendations for reform are made in Part 4 prior to conclusion of the chapter in Part 5.

Engaging private sector finance in climate governance

The business case and overcoming barriers to private sector engagement

Private sector actors make investment and financing decisions based on 'the business case', which comprises both hard (financial) and soft (reputational) factors. Making a business case comprises a cost/benefit analysis, being an assessment of projected returns and potential risks inherent to a particular investment or project. The upshot is that if an initiative is not profitable or if the risks outweigh potential benefits – that is, if there is *not* a business case to do it – then private sector actors, such as corporations and institutional investors, will not get on board.

Mobilizing private finance for mitigation and adaptation thus requires overcoming certain barriers to engagement due to investor reticence. Reasons for this include (Bowman 2016): basic unawareness of climate risk; long time-frames for major infrastructure which require long-term risk/benefit analyses into the future but for which there is information asymmetry due to information gaps; projects may provide 'social returns' (or community benefits) but not financial returns, which is particularly salient for adaptation projects; 'know how' deficit regarding climate-related financing in addition to high transaction costs associated with novel modalities of financing; and political and sovereign risk (perceived or actual) especially in jurisdictions with fluctuating regulation.

As such, overcoming barriers and re-balancing the business case in favour of private sector engagement requires government intervention and support. Law and policy can modify the risks and returns faced by private sector actors as well as the information and processes they use for their decision-making. This, in turn, influences whether and to what extent finance will flow to climate change mitigation and adaptation endeavours at the necessary scale.

Financial and facilitative regulation for public–private climate finance

When endeavouring to harness business case logic and overcome barriers to private sector engagement, regulation needs to incentivize private sector actors to finance and invest in greener initiatives (Bowman 2015). This will take various forms, including blended financing to incentivize engagement through concessional arrangements; guarantees and use of insurance to reduce investment risks; green bonds; tax credits; grants and subsidies; a carbon price (tax or trade). We can call these *Financial Mechanisms*, being 'hard' financial instruments that directly mobilize or leverage private finance through economic incentives and disincentives that nudge action in a particular direction. Much of the climate finance literature and government policy responses tend to focus on this category.

Yet my recent analysis of international climate finance case studies reveals that non-financial and informational modalities are just as important as financial mechanisms for ensuring flows of private capital (Bowman 2016). I term these mechanisms as *Facilitative Modalities*, being 'soft' non-financial measures that compliment hard Financial Mechanisms for mobilizing greener private capital by enhancing project pipeline, capacity building and knowledge transfer. Examples of Facilitative Modalities are government support for creation of centralized information repositories; supporting corporate and financial disclosures for a 2 degree Celsius (2DC) transition; tracking finance for climate-related purposes and measuring outcomes and impacts; and knowledge sharing and capacity building through 'matchmaking & training' platforms and idea labs. This last grouping can be highly effective for overcoming barriers of lack of awareness, information or know-how in order to align investable projects with capital streams. An admirable example is the Private Financing Advisory Network (PFAN), which was created within the Climate Technology Initiative of the International Energy Agency: it identifies promising clean energy projects at an early stage and provides mentoring to improve the chances of implementation by hiring expert consultants to advise project proponents about developing and making a business case for their project and also by working with local investors to help educate them about these new types of projects. In its early years, the PFAN had mobilized many millions in private finance which cost only 5% in public money to stimulate.

In short, an *integrated regulatory framework* comprising both financial mechanisms and facilitative modalities is required to successfully and sustainably mobilize, leverage and scale-up private climate finance.

The European Fund for Strategic Investment (EFSI) and climate finance: critiques and challenges

EFSI purpose and progress

The EFSI is core to the Investment Plan for Europe and is a facility aimed at promoting private investment throughout the EU by a range of risk mitigating and co-financing instruments such as guarantees, loans, capital market instruments and equity participation (MEMO/15/5419). Its foundation is 'the EU Guarantee' which is an 'irrevocable and unconditional guarantee by the Union to the EIB for financing and investments operations in accordance with the EFSI Regulation' (EFSI Regulation arts 8, 9, 11). The EIB Group (comprising the EIB and European Investment Fund (EIF)) can then provide a first loss guarantee to indemnify investors for a given amount of any losses. The EFSI is split into two windows: the SME Window; and the Innovation and Infrastructure Window. The former targets Small-Medium Enterprises and is governed by the EIF through financial intermediaries (Agricultural Markets Task Force 2016); it was particularly successful in the first year of implementation (European Commission 2016b). The Innovation and Infrastructure window (IIW) is managed by the EIB and is the main EFSI investment vehicle that is the focus of much of this chapter.

Moreover, in conjunction with the EFSI, the EIAH of the Investment Plan for Europe aims to enable investment by providing access for investors to advisory and technical assistance programmes. It develops risk profiles to ensure private actors have a comprehensive knowledge of a sector in which they might otherwise be hesitant to invest due to a lack of experience or know-how (European Commission n.d., pp. 13–14; European Commission and European Investment Bank n.d.). By combining the EFSI and EIAH, the Investment Plan is well-placed to mobilize finance from conservative actors, such as pension funds and banks, which manage and allocate vast amounts of money but are risk-averse (European Investment Bank n.d.).

Indeed, the EFSI has been successful in drawing private sector attention to the need for increased investment into Europe (Business Europe 2016, p. 4). The initial endowment of the EFSI was €21 billion (comprising the €16 billion guarantee from the EU budget with €5 billion allocation of the EIB's own capital) to equip the EIB to borrow €61 billion to invest in projects from 2014–2017 (European Commission 2014, p. 2). This is expected to trigger €315 billion in additional investment by mid-2018 (MEMO/15/5419). The scheme so far has been labelled by the EIB a success, unlocking €116 billion in private funds in the first 12 months following its establishment (European Investment Bank 2016, p. 6). However, the implementation of the scheme has been criticized in part because the overall investment gap in Europe, which the fund was designed to close, requires an estimated injection of between €200 billion and €300 billion per year (Rubio, Rinaldi and Pellerin-Carlin 2016, p. 12). Therefore, although a sizeable step, the €116 billion does not actually close the required gap.

Thus, in 2016, the EC presented a proposal for *EFSI 2.0*, which involves an extension 'to double the duration of the Fund and double its financial capacity to provide a total of at least €500 billion of investments by 2020' (Juncker 2016). This would be achieved by increasing the EC guarantee to 26 billion and increasing the capital contribution from the EIB to 7.5 billion (Council of the European Union 2016). The proposal was approved by Members of the European Parliament's Budgets and Economic and Monetary Affairs committee on 24 May, 2017 and, at the time of writing, was awaiting approval by the

European Parliament and the Parliaments of the Member States (European Commission 2017a).

As the EFSI heads into its second phase, this is a pivotal juncture to analyse how it could be improved and extended not just economically but also environmentally and inclusively. In order to recommend improvements in the next Part, the remainder of this Part will analyse its performance regarding facilitating green and climate-friendly EU ambitions.

Hierarchy of needs? Prioritizing green projects and conflicting goals

The eligibility criteria for project approval are set out in Article 6(1) of the EFSI Regulation (Regulation (EU) 2015/1017). In short, EFSI projects must (1) be economically viable; (2) be consistent with Union policies; (3) carry an element of 'additionality'; (4) maximize, where possible, the utilization of private sector capital; and (5) be technically viable. While there is no overt reference to climate-friendly investment for greener projects or infrastructure, article 6(1)(b) mentions the Union 'objective of smart, sustainable and inclusive growth', which engages with the 2030 Framework and has potential to facilitate Paris Agreement objectives. Indeed, to date, about half of the approvals for IIW have involved energy efficiency or climate-friendly projects (Cañete 2016) with examples including *Copenhagen Infrastructure II*, an infrastructure fund investing in large wind and biomass energy-related projects and transmission; the *Spark Project* involving the installation of smart-meters across the United Kingdom; and increasing energy efficiency in buildings in France (MEMO/16/2195, pp. 3–5).

However, the broad nature of the eligibility criteria when read together with EU policies and EFSI objectives in art. 9(2) EFSI Regulation means that many different types of projects can be funded whether climate-friendly or not. The issue is that no hierarchy or prioritization is provided. Thus, project approvals can create conflict between competing EU policy objectives. For example, jobs, growth and infrastructure investment are high priorities in Juncker's Commission (European Commission n.d., p. 5), yet those areas are not inherently climate-friendly and in fact can be achieved without any enhanced attention to sustainability. Indeed, a recent assessment by the Climate Action Network (CAN) Europe of EFSI funding found that 15% of its energy sector financing was for fossil fuel-related projects, and that 68% of funding in the transport sector supported high-carbon projects (such as motorways and airports) (Climate Action Network Europe 2016, p. 2).

This lack of hierarchy is not only confusing but also likely to undermine progress made on climate and sustainability-related EU policy objectives, particularly the ambitious 40% reduction in GHG emissions by 2030. There is a clear issue here of long-term vs. short-term thinking. Investing in carbon-heavy projects is unsustainable in not only environmental terms but also fiscal ones. Europe's 2050 goals for Europe are to phase out GHG emissions (aiming for an 80–95% overall reduction) (European Commission 2017c); yet according to a 2016 Oxford University report, the 2DC guardrail requires that no new GHG-emitting electricity infrastructure can be built or financed after 2017 (Pfeiffer, Millar, Hepburn and Beinhocker 2016). Therefore, the most economically prudent course of action is to shift all new energy investment to zero carbon *from now*.

Geographical coverage and mitigation focus

By seeking to decrease risk for private actors, the EFSI first loss guarantee seems primarily aimed at reduction of sovereign risk and regulatory risk associated with fluctuating government policy. Therefore, markets with proven records of regulatory certainty are more appealing for institutional investors and, for this very reason, investments into these jurisdictions are less likely to need public financial support. As such, there is a clear role for the EFSI to incentivize private sector engagement in countries with higher risk (perceived or actual) and especially in climate-related projects. The *Euromoney Country Risk Analysis* shows that such states are generally Southern and Eastern European nations (Norton Rose Fullbright, 2014); with Poland and the Czech Republic recently attempting to block EU climate-friendly initiatives to kick start a 'prosperous' economy based on fossil fuels (Nelson 2017). To date, however, the EFSI has not performed this important role (Business Europe 2016, pp. 4–5). Current IIW figures show that 52% of projects were approved in just three states being the United Kingdom, Italy and France, with only 12% of projects approved in central European countries (Climate Action Network Europe 2016, p. 3).

Further, EFSI funding for IIW projects has been allocated to climate mitigation with little attention to adaptation. This is consistent with EU policy which emphasizes a low-carbon transition; but it does not facilitate Paris Agreement goals as outlined in the introduction which give attention to financial assistance for both mitigation and adaptation, and especially for developing countries. For conservative investors like pension funds, the long-term nature of energy and infrastructure projects and the untested nature of some green technologies raise the concern that risk may transpire after the project is completed but before return materializes (Bowman 2016, pp. 7–8). This is particularly relevant with adaptation measures that have predominantly social (not financial) returns, as outlined in Part 2. Despite this fact, approximately 30% of IIW investment has occurred in energy efficiency and renewable energy sectors in strong economies (Climate Action Network Europe 2014, p. 2). Importantly, this support was focused in developed countries: 70% of renewable energy finance took place in Belgium; 80% of energy efficiency investments occurred in France, Finland and Germany (*Bankwatch Briefing* 2017, p. 2). Looking forward, this focus on mitigation projects in developed countries is problematic for achieving not only Paris Agreement goals but also EFSI goals of inclusivity and additionality. On this latter point, it is worth noting that solar and wind projects reached parity with fossil fuels in over 30 countries in 2016 with the World Economic Forum describing these forms of renewable energy as 'an outright compelling investment opportunity with long-term, stable, inflation-protected returns'. (World Economic Forum 2017; Griffin 2017). In short, they are becoming commercially viable and require less government support, especially in developed countries.

Thus, there is a clear need for the EFSI to facilitate funding for: first, adaptation projects Europe-wide and, second, mitigation projects in jurisdictions with which investors have less transaction experience and which may have higher sovereign risk or less developed financial sectors, as seen in the examples of Poland and the Czech Republic. Indeed, the EIB Vice President has acknowledged that engaging private actors in Eastern European States is more difficult due to their less developed financial sectors, which therefore decreases investment confidence (Barbière 2016). The result is an investment bias 'towards those countries with the largest supplies of long-term lending resources, meaning the most wealthy – and those with the most extensive finance sectors' (Myant 2015). This bias is something the EFSI should actively work to counteract rather than consolidate.

Additionality

Concerns regarding geographical coverage and adaptation extend into that of additionality. To engage the EFSI, projects require an element of 'additionality', which describes projects that could not be carried out at all or to the same extent without EFSI support (Art. 5, EFSI Regulation). Adapting this criterion to Paris Agreement objectives, there is a key role for the EFSI to promote investment in mitigation and adaptation climate-related projects that are inherently riskier or have lower fiscal returns, and which therefore would not naturally attract large-scale private investment.

However, in practice, the 'additionality' required under EFSI projects is vague and not subject to a regulatory framework: the EFSI being but a tool to increase investment, and not a policy itself (Business Europe 2016, p. 4). So, for example, it does not require attention to the innovative value of projects (CAN Europe 2016, p. 4) nor does it focus on projects that could produce some greater benefit for the EU beyond increased investment (Counter Balance 2016).

Indeed, one study has found that 42 out of 55 projects funded in the IIW were extremely similar in description to projects funded by the EIB prior to the EFSI (Claeys and Leandro 2016, noting however their methodology of basing similarity purely on topic description and not on financials). Moreover, approved projects may meet the 'additionality' criteria only because the guarantee creates more risk for the EC, thus putting it and the EIB in a subordinated position, rather than the projects themselves carrying more risk (being the types of projects for which the fund was intended) (S&P Global Ratings 2017, p. 10). This could create an undesirable situation for the EU by privatizing profits and socializing losses (De Masi 2016). Relatedly, there is concern that the eligibility criteria of maximizing utilization of private sector capital may result in 'de-politicized' funding decisions that are designed purely to maximize profitability rather than EU overall wellbeing (*GUE/NGL Discussion Paper* n.d.). Since the advent of the Paris Agreement and in light of the EU 2030 goals, notions of 'wellbeing' now include de-carbonization of the economy. Thus, a balance between the two aims of enhancing profitability via business case logic (for the private sector) and enhancing wellbeing (for the public) needs to be achieved.

In short, the way projects are assessed as 'additional' at present does not necessarily favour climate-friendly projects generally nor distinguish between mitigation and adaptation specifically as discussed earlier. One could even envisage a situation where carbon-heavy projects, as they become phased out and therefore riskier investments, become more compatible with the additionality criterion for EFSI funding even as they become less desirable. The effect is that the EFSI would simply extend the reach of the EIB into riskier projects rather than leveraging public funds to further EU goals by addressing climate change and enhancing sustainable finance (S&P Global Ratings 2017, p. 10).

Ways forward: recommendations and reform

What instruments or changes can combat the particular shortcomings of EFSI identified in Part 3 and in light of general investment barriers identified in Part 2? How might the EFSI be reformed to incentivize greener private sector engagement to implement Paris Agreement objectives and enhance EU wellbeing?

(Re)Prioritizing and additionality

With a view to combatting additionality concerns, the *EFSI 2.0* proposal has identified the need to address sub-optimal investments and market gaps (MEMO/16/2983). However, the guidance at this stage is vague and lacks significant strategy for how this might occur optimally. Specifically, it reserves 40% of funds in the IIW for projects aligned with meeting objectives under the Paris Agreement and the EU's self-set goals in climate change (ibid.); yet this raises questions of how that percentage is to be achieved and whether it is sufficient. In particular, there is an obvious concern that the remaining 60% of funds are available for climate-*unfriendly* projects. Indeed, under the new proposal, cross-border infrastructure (including high carbon motorway extensions) would be considered 'very additional' and thereby would almost always come under the scheme (Climate Action Network Europe 2016, p. 4). This is clearly undesirable in the context of Paris Agreement goals and the EU 2030 Framework.

Significant changes ought to be made regarding transparency in determining how and to what extent projects match the eligibility criteria especially the 'additionality' requirement. A key underlying issue is a lack of transparency in how these projects are assessed as 'additional' given that the EIB is not obliged to publish its reasons for providing funding (Claeys and Leandro 2016; Counter Balance 2016) or to justify how an approved project is compatible with EU policy. The Independent Investment Committee, which is responsible for approving projects, should be required to justify its funding decisions (European Parliament 2017), including publishing a scoreboard of indicators, such as how exactly projects meet Union policy objectives and additionality.

Moreover, the EFSI could look to the recently formed Green Climate Fund (GCF) for inspiration regarding coordination between relevant EU actors. The GCF Board is collaborating with other relevant UNFCCC entities, such as the

Adaptation Committee and the Technology Executive Committee, to define and enhance linkages between them (Decision 13/CP.21, 2016). The aim is to act in concert to de-risk and scale-up private investments, particularly in areas such as climate technology projects and programmes that are relatively unknown terrain for large, conservative investors (Minas and Bowman 2017). Mirroring this approach, the EFSI ought to collaborate with other relevant EU Funds, to further reduce perceived and actual risks associated with climate friendly projects. Such Funds include the European Structural and Investment Fund, the Private Finance for Energy Efficiency instrument, and the Connecting Europe Facility (a fund that supports the development of high performing, sustainable and efficiently interconnected trans-European networks) (European Parliament 2017).

The HLEG on Sustainable Finance had suggested introducing an explicit ban on fossil fuels and unsustainable projects (High-Level Expert Group on Sustainable Finance 2017, p. 45). It argued that this would prevent the de-risking of unsustainable investment and thus maximize the effectiveness of public funds in mobilizing private finance by preventing the inefficient allocation of capital in stranded assets and projects with a clear expiry date.

Geographical inclusivity

To redress the currently limited geographical scope of the fund, some have called for the introduction of geographical quotas to ensure poorer regions have access to the scheme (European Alliance 2016). However, industry players have not embraced that idea (Business Europe 2016, p. 5), arguing that a strength of the scheme is its flexibility and demand driven nature (European Commission n.d., p. 11). An alternate approach is to make local information available to investors to instil confidence and encourage investment in less-advantaged member states (Griffith-Jones, n.d.).

Specifically, this requires increased attention to facilitative modalities in EFSI and EIAH activities. To date, project approvals have utilized financial mechanisms, especially for mitigation and particularly in developed countries. However, as explained in Part 2, facilitative modalities are key to massaging project pipeline, building capacity and enabling knowledge sharing. The EIAH could be far better utilized in this regard to fulfil the aims of the Investment Plan.

Under *EFSI 2.0*, the EC proposes to strengthen the role of the EIAH in less developed regions, and provide greater opportunity for National Promotional Banks (NPBs) to act as co-financers and provide risk assessments. The aim is to decentralize information platforms and equip the EIAH with specialized local knowledge to tailor projects for private sector investment in priority areas (European Commission n.d., p. 14).

In contrast, the HLEG on Sustainable Finance recommended creating a new organization 'Sustainable Infrastructure Europe' to act as a matchmaking facility that also develops and structures infrastructure projects with a strong business case (High-Level Expert Group on Sustainable Finance 2018, pp. 34–38). Doing so would, happily, model the EIAH as a PFAN-style entity.

Whether the EIAH continues as an expanded entity or is leveraged by a new institution, giving it domain over facilitative modalities would help to create more project pipeline, enhance capacity in nations that need it most and mitigate perceived investor risks stemming from a lack of know-how and familiarity. In addition, adopting a GCF-inspired 'linkage' approach with the EFSI and other relevant EU Funds, as outlined earlier, would help to maximize each Fund's specialist knowledge of their particular sectors to massage know-how sharing and project pipeline and help equip a revamped EIAH with local knowledge and information. This in turn could provide a truly powerful information platform to scale-up private investment in both mitigation and adaptation projects in a broader variety of EU member states. In addition, the HLEG on Sustainable Finance has recommended creating a 'sustainability taxonomy' as a 'shared EU classification system for sustainable activities', which could help private actors to familiarize themselves with the minimum protections afforded in all Member States and thus mitigate sovereign risk (High-Level Expert Group on Sustainable Finance 2018, pp. 10, 36).

Finally, the EFSI could better utilize its finance for SMEs to facilitate a green transition. Currently, the SME Window faces the same issue as the IIW whereby the focus is on investment generally and not aligned with the EU's overall goals (European Investment Bank n.d.). Yet SMEs comprise 99 out of 100 businesses in the EU (European Banking Authority 2015, p. 9) and, as a major driver of innovation and employment, they are vital in the transition to a sustainable economy (McDaniels and Robins 2017, pp. 11, 13). The HLEG interim recommendations did not include mention of SMEs; yet the EU is implicitly aware of their importance due to programmes such as InnovFin, which facilitate access to the Horizon 2020 scheme by innovative but high-risk SMEs (ibid., p. 24). Given the significance of SMEs' role in a sustainable society, it is important to not only ensure that the IIW is directed towards financing a sustainable economy, but also to make sure SMEs that strive to work towards green goals are rewarded with access to a wider range of funding options through the SME Window.

Conclusion

The EU seeks to position itself as a global leader for climate change diplomacy and action, and it also seeks to rebuild itself through enhanced internal investment. These twin aims share an important commonality: engaging the private sector for public purpose.

In July 2017, Christian Thimann stated that: 'The EU has been leading on the agenda for making the transition to a sustainable European economy and a sustainable financial system. The challenge now is to develop a clear strategy that unifies this ambition with one of the EU's key achievements in terms of financial policy and regulation, and which funds the path towards a low-carbon, resource-efficient and environmentally protective economy' (Thimann 2017, p. 2).

This chapter has responded to that challenge by arguing that the Investment Plan for Europe, through both investment and informational elements, could provide an effective financial and regulatory vehicle through which the EC can mobilize climate-related investment to enhance both EU regional implementation and international leadership of Paris Agreement goals. In so doing, it could also enhance a sustainable financial system. Indeed, in a post-Paris world, 'sustainable finance' must include not only inclusive finance for resource efficiency and low-carbon goals, but also capital for adaptation and climate-resilient development especially in developing countries as well as increased attention to the intermediating role of SMEs. These points deserve explicit attention in both EU diplomacy and policy.

Looking forward, the *EFSI 2.0* proposal seeks to introduce an EFSI-type scheme to developing countries abroad, such as Africa, in order to mobilize private finance and channel private investment into infrastructure and SMEs. To do so, the EC proposes establishing an External Investment Fund to 'offer credit enhancement, risk sharing, investment grants and technical assistance based on a guarantee and funding from the EDF/EU budget and other contributors'. (COM(2016) 385 final, p. 12). This initiative is laudable. However, this chapter has demonstrated that in order to achieve status as an international role model for implementing the Paris Agreement and facilitating sustainable finance, the EC needs to pay more attention to the shortcomings and potential reforms for climate-friendly investment on its home shores.

Note

1 I am very grateful to research fellow Daniel Maglieri for his invaluable assistance through the King's Undergraduate Research Fellowships scheme.

References

Agricultural Markets Task Force. (2016). "Issue Paper: The European Fund for Strategic Investment (EFSI) – The Investment Plan for Europe." 12 April 2016.

Bankwatch Briefing. (2017). "The Winners and Losers of Climate Action at the European Investment Bank" May 2017. https://bankwatch.org/sites/default/files/Briefing-EIB-climateaction-May2017.pdf (accessed 8 June 2017).

Barbière, C. (2016). "Juncker Plan Struggles to Make Inroads in Eastern Europe." *EURACTIV* 25 May 2016. www.euractiv.com/section/euro-finance/news/juncker-plan-struggles-to-make-inroads-in-eastern-europe/ (accessed 15 June 2017).

Bowman, M. (2015). *Banking on Climate Change: How Finance Actors and Transnational Regulatory Regimes are Responding*. Kluwer, Dordrecht, The Netherlands.

Bowman, M. (August 2016). "Innovative International Climate Finance Approaches." *Final Options Paper for the Australian Department of Foreign Affairs and Trade*.

Business Europe. (November 2016). "Extending the European Fund for Strategic Investment." *Investment Plan for Europe: Are We There Yet?*

Cañete, M.A. (13 June 2016). "EFSI can Help EU Realise Its COP21 Ambitions," *The Parliament Magazine*. www.theparliamentmagazine.eu/articles/opinion/efsi-can-help-eu-realise-its-cop21-ambitions (accessed 7 June 2017).

Claeys, G. and Leandro, A. (17 May 2016). "Assessing the Juncker Plan after One Year." http://bruegel.org/2016/05/assessing-the-juncker-plan-after-one-year/ (accessed 6 June 2017).
Climate Action Network Europe. (October 2016). "EFSI 2.0: Make it Climate Proof, Additional and Transparent." *Briefing on the Legislative Review of the European Fund for Strategic Investments*.
COM(2014) 903 final. "An Investment Plan for Europe." *Communication from The Commission to The European Parliament, The European Council, The European Central Bank, The European Economic and Social Committee, The Committee of the Regions and The European Investment Bank* 26 November 2014, Brussels, Belgium.
COM(2016) 385 final. "Establishing a New Partnership Framework with Third Countries Under the European Agenda on Migration." *Communication from The Commission to The European Parliament, The European Council, The Council and The European Investment Bank* 7 June 2016, Strasbourg, France.
COP22, final. (2016). "Marrakech Partnership for Global Climate Action."
Council of the European Union. (6 December 2016). "Investment Plan for Europe: EFSI Extension Agreed by Council."
Counter Balance. (January 2016). *Briefing for Board of Directors for the Seminar with Civil Society on 1st February 2016: European Fund for Strategic Investments – How to Ensure Sustainability, Good Governance and Added-Value of EFSI Operations*. www.counter-balance.org/wp-content/uploads/2016/01/EFSI-Briefing_for-EIB-Board.pdf (accessed 7 October 2017).
De Masi, F. (28 January 2016). "The Juncker Plan: Failing to Address the Investment Gap." Brussels, Belgium.
Decision 13/CP.21. Linkages between the Technology Mechanism and the Financial Mechanism of the Convention, FCCC/CP/2015/10/Add.2, 29 January 2016.
EFSI Regulation, Regulation (EU) 2015/1017 of The European Parliament and of The Council of 25 June 2015 on the European Fund for Strategic Investments, the European Investment Advisory Hub and the European Investment Project Portal and amending Regulations (EU) No 1291/2013 and (EU) No 1316/2013.
European Alliance. (29 June 2016). "Szwabski Calls for Geographical Quota for Poorer Regions to Have Access to EU Investment Fund." http://web.cor.europa.eu/ea/News/Pages/Szwabski-calls-for-geographical-quota-for-poorer-regions-to-have-access-to-EU-investment-fund.aspx# (accessed 1 August 2017).
European Banking Authority. (2015). "EBA Discussion Paper and Call for Evidence on SMEs and the SME Supporting Factor" EBA/DP/2015/02 31 July 2015.
European Commission. (24 November 2014). "Fact Sheet: Where Does the Money Come From?"
European Commission. (22 December 2016a). "Press Release: European Commission Appoints Members of the High-Level Expert Group on Sustainable Finance." Brussels, Belgium. http://europa.eu/rapid/press-release_IP-16-4502_en.htm (accessed 5 June 2017).
European Commission. (1 June 2016b). "Press Release: Investment Plan for Europe: European Fund for Strategic Investments to be Extended Following Successful First Year." http://europa.eu/rapid/press-release_IP-16-1933_en.htm (accessed 17 June 2016).
European Commission. (24 May 2017a). "Investment Plan: Commission Welcomes Vote by European Parliament Committees as EFSI 2.0 Nears Final Adoption." https://ec.europa.eu/newsroom/ecfin/item-detail.cfm?item_id=66932&utm_source=ecfin_newsroom&utm_medium=Website&utm_campaign=ecfin&utm_content=Investment%20Plan:%20Commission%20welcomes%20vote%20by%20European%20Parliament%20committees&lang=en (accessed 7 June 2017).

European Commission. (2017b). "Paris Agreement." *Climate Action*. https://ec.europa.eu/clima/policies/international/negotiations/paris_en (accessed 5 June 2017).

European Commission. (2017c). "EU Climate Action." *Climate Action*. https://ec.europa.eu/clima/citizens/eu_en (accessed 5 June 2017).

European Commission. (n.d.). "The Investment Plan for Europe: State of Play."

European Commission and European Investment Bank. (n.d.). "European Investment Advisory Hub." https://ec.europa.eu/inea/sites/inea/files/10b.cef_ener_20160407_ecfin_eiah.pdf (accessed 7 June 2017).

European Investment Bank. (4 December 2015). "Low Level Draws High Investment." www.eib.org/infocentre/stories/all/2015-december-04/low-leverage-draws-high-investment (accessed 9 June 2017).

European Investment Bank. (2016). "Investment Plan for Europe: The First Year."

European Parliament. (January 2015). "Briefing: EU Position for COP 21 Climate Change Conference."

European Parliament. (4 May 2017). "European Fund for Strategic Investments – EFSI 2.0." *Briefing: EU legislation in progress*.

Godinot, S. (8 June 2017). "EU Commission Paves Way for Ambitious Strategy on Sustainable Finance." www.wwf.eu/?302110/EU-Commission-paves-way-for-ambitious-strategy-on-sustainable-finance (accessed 14 June 2017).

Green Climate Fund. (n.d.a). "Private Sector." www.greenclimate.fund/what-we-do/private-sector-facility (accessed 14 June 2017).

Green Climate Fund. (n.d.b) "Annex III: Initial Investment Framework: Activity-Specific Sub-Criteria and Indicative Assessment Factors" GCF/B.09/23. www.greenclimate.fund/documents/20182/239759/Investment_Criteria.pdf/771ca88e-6cf2-469d-98e8-78be2b980940 (accessed 1 August 2017).

Griffin, A. (4 January 2017). 'Solar and Wind Power Cheaper than Fossil Fuels for the First Time." *The Independent*. www.independent.co.uk/environment/solar-and-wind-power-cheaper-than-fossil-fuels-for-the-first-time-a7509251.html (accessed 30 June 2017).

Griffith-Jones, S. (n.d.). "4.1 Enhance the Role of the EIB and National Development Banks," *EFSI, brief evaluation*. www.progressiveeconomy.eu/sites/default/files/EFSI.pdf (accessed 6 June 2017).

GUE/NGL Discussion Paper. (n.d.). "What Europe Needs: A True Public Investment Plan."

High-Level Expert Group on Sustainable Finance. (July 2017). "Financing a Sustainable European Economy." *Interim Report*.

High-Level Expert Group on Sustainable Finance. (January 2018). "Financing a Sustainable European Economy." *Final Report*.

Holmes, (27 February 2017). "What's Next for Sustainable Finance in Europe? An E3G View." www.e3g.org/library/what-next-for-sustainable-finance-in-europe-an-e3g-view (accessed 14 June 2017).

International Energy Agency. (2014). "Special Report: World Energy Investment Outlook." OECD/IEA 2014.

Juncker, J. (14 September 2016). "State of the Union Address European Parliament."

Latvia and European Commission. (6 March 2015). "Intended Nationally Determined Contribution of the EU and its Member States." *Submission by Latvia and the European Commission on behalf of the European Union and its Member States*.

McDaniels, J. and Robins, N. (2017). "Mobilising Sustainable Finance for Small and Medium Sized Enterprises: Reviewing Experience and Identifying Options in the G7." *Enquiry: Design of a Sustainable Financial System*. United Nations Environment Programme.

MEMO/15/5419, European Commission "Fact Sheet: The Investment Plan for Europe: Questions and Answers." 20 July 2015.

MEMO/16/2195, European Commission "Fact Sheet: The Investment Plan for Europe and Energy: Making the Energy Union a Reality." 14 June 2016, Brussels, Belgium.

MEMO/16/2983, European Commission "Fact Sheet: State of the Union 2016: The European Fund for Strategic Investments (EFSI) – Frequently Asked Questions" 14 September 2016, Strasbourg.

Minas, S. and Bowman, M. (2017). "Post-Paris/Post-Trump: The Green Climate Fund and Climate Finance Governance in the Eye of the Storm." *TLI Think! Paper 75/2017; Peking University School of Transnational Law Research Paper No. 17–6*. https://papers.ssrn.com/sol3/papers.cfm?abstract_id=3038512 (accessed 7 October 2017).

Myant, M. (29 January 2015). "Why Juncker's Investment Plan is a Good Try but Not Enough." *Social Europe*. www.socialeurope.eu/2015/01/junckers-investment-plan-good-try-not-enough/ (accessed 15 June 2017).

Nelson, A. (29 May 2017). "EU Climate Laws Undermined by Polish and Czech Revolt, Documents Reveal" *Climate Home*. www.climatechangenews.com/2017/05/29/eu-climate-targets-undermined-polish-czech-revolt-documents-reveal/ (accessed 15 June 2017).

Norton Rose Fullbright. (December 2014). "European Infrastructure Opportunities: An Investment Plan for Europe."

Pfeiffer, A., Millar, R., Hepburn, C. and Beinhocker, E. (24 March 2016). "The '2°C Capital Stock' for Electricity Generation: Committed Cumulative Carbon Emissions from the Electricity Generation Sector and the Transition to a Green Economy." *Applied Energy*. www.sciencedirect.com/science/article/pii/S0306261916302495 (accessed 1 August 2017).

Rubio, E., Rinaldi, D. and Pellerin-Carlin, T. (March 2016). "The Investment Plan for Europe: Making the Best of the Juncker Plan." *Jacques Delors Institute, Studies & Reports 109*.

S&P Global Ratings. (20 March 2017). "Europe's Investment Plan Surges to €500 billion, but is it Working?" www.spratings.com/documents/20184/1481001/Europes+Investment+Plan+Surges+To+500+Billion+Euros+But+Is+It+Working+Correction2/fd3bc20a-74ae-49bd-85fb-16292ff707bd (accessed 7 June 2017).

Thimann, C. (18 July 2017). "The Ambition and Challenges of Sustainable Finance: Presentation of the Interim Report." *Speech at the Stakeholder Conference organised by the European Commission*. Brussels, Belgium.

United Nations. (12 December 2015). "Decision of Parties to Adopt Paris Agreement: Proposal by the President – draft decision CP/.21." FCCC/CP/2015/L.9.

Walsh, J. (3 April 2017). "European Commission's Sustainable Finance Committee Targets Investment Chain Transformation." *Environmental Finance*. www.environmental-finance.com/content/news/european-commissions-sustainable-finance-committee-targets-investment-chain-transformation.html (accessed 14 June 2017).

Walsh, J. and Cooper, G. (14 July 2017). "Experts Call for Reform of Financial Regulation to Promote Sustainability. *Environmental Finance*. www.environmental-finance.com/content/news/experts-call-for-reform-of-financial-regulation-to-promote-sustainability.html?utm_source=170717na&utm_medium=email&utm_campaign=alert (accessed 1 August 2017).

World Economic Forum. (2017). *Renewable Infrastructure Investment Handbook: A Guide for Institutional Investors*.

Section 4

Negotiations

9 Climate technology diplomacy through the UNFCCC and beyond

Matthew Kennedy

Introduction

Governments face global environmental and security challenges as they strive to reduce fossil dependence and achieve affordable energy to meet their economic aspirations. Global warming, carbon emission penalties and finite fossil fuel reserves have all contributed to a surge in climate technology deployment, pointing the way towards surmounting these challenges and achieving a more sustainable future. Governments and policymakers are being urgently asked to act to reverse emissions growth trends and let scientific evidence inform their pathway. There is international acceptance that a response to tackling climate change requires international diplomacy and cooperation on a global scale.

With the OECD (2015) estimating that CO_2 emissions are estimated to reach around three times more than what would be required to limit the long-term increase in global temperatures to two degrees, an accelerated investment in climate technology deployment and technology transfer is more urgent than ever. While technological diffusion remains difficult to achieve, identifying the appropriate conditions and environment to strengthen technological co-operation and aid technology transfer to developing countries is becoming critical. The focus of this chapter is on the institutional and diplomatic linkages that can foster international diplomacy co-operation to enable technical capacity to be built in development countries and so enable the deployment of technology at scale. While the key area of focus is the United Nations Framework Convention on Climate Change (UNFCCC), the paper will introduce additional co-operative mechanisms including the International Energy Agency (IEA), International Renewable Energy Agency (IRENA) and Mission Innovation.

Climate diplomacy and collaboration

The UNFCCC stresses the priority economic and sustainable development needs of developing countries, as energy access concerns and associated poverty engulf countries. The provision of clean energy services to society, while minimizing health risks and protecting the environment, combines to form part of the wider climate change agenda. Countries ('parties', in the language of the Convention) agree that diplomacy and co-operation across finance, technology and capacity building is required to meet stabilization objectives. Prior to COP21, the climate change regime focused on 'what' actions need to happen across its framework of targets, plans and objectives, rather than elaborating and enabling 'how' action happens. The Paris Agreement provided a collective commitment to avoid emissions growth with stated goals in terms of peak emissions reflected in more precise policy actions. It introduced the rules and targets for climate stabilization and emissions reduction in terms of the required emissions trajectory. It outlined countries' pursuit of domestic measures as expressed in the nationally determined contributions (NDC).[1] Almost 200 parties drew on lessons, principles and the institutions required to bring this climate transition and introduced both flexibility and robustness in terms of the 195 global NDCs.

The diplomatic efforts concerning the framing of the Paris Agreement reflected the architectural complexity in terms of the setting and achievement of carbon dioxide emissions reduction targets, expressed in terms of sustainable development, energy security and environmental protection. This became visible in terms of a country's investment decisions for a premature retirement of energy technology infrastructure or a mass deployment of climate technology. While technology development and transfer can enable the achievement of the mitigation and adaptation provisions outlined under the UNFCCC, it became clear, through submitted NDCs, that there is a limited understanding of the variables that influence a wider adoption of such technologies and the associated opportunities. When one interprets the global analysis required to aid the low carbon transition, it becomes clear that many of the least developing countries (LDCs) and small island developing nations (SIDS) lack the modelling tools and technical architecture to plan such transitions. This lack of capacity results in parties seeking flexibility within their national commitments, a flexibility that takes into account their political and physical circumstances, capabilities and legal systems.

One of the main challenges to successful climate technology deployment involves the identification of new ways to widely deploy existing technologies or to adapt new technologies to local conditions (WBCSD 2010). Global emissions growth and demand trajectories, expressed in million tonnes of oil equivalent, are projected to occur mostly in China, the Middle East, Brazil and India, and curbing this growth will require co-operation and diplomacy that re-orientates the energy sector. This is most likely achieved through the promotion, diffusion and transfer of climate change technology for mitigation and adaptation.[2] The importance of strengthening of endogenous technological capabilities within these countries, while fuelling economic growth and social progress and avoiding fossil lock-in, is often underestimated.

The emergence of climate technology under the UNFCCC

As technology largely determines the level of emissions, it is both the primary source of the climate change problem and the key to the solution (Carpenter 2012). Climate technologies have the ability to meet emerging societal challenges that range from enhancing industrial competitiveness to limiting vulnerabilities associated with fossil dependence and addressing stabilization. As countries agreed to collaborate on a set of technology transfer activities under the UNFCCC Marrakesh Accords (2001, Article 4.5), it was the Kyoto Protocol's Clean Development Mechanism that indirectly provided the opportunity to gain access to advanced climate technologies and low carbon infrastructure in developing countries. While the CDM indirectly provides the opportunity to gain access to advanced technologies and low carbon infrastructure to developing countries, much of the mobilized investment has gone to developed countries with limited technology transfer (Nicholls 2008) and with questionable success (Haites et al. 2006; de Coninck et al. 2007).

Climate technology fully emerged as one of four 'building blocks' towards a binding agreement cited within the UNFCCC's Bali Action Plan (2007, Para 1.d), alongside mitigation, adaptation and finance. While COP15 in Copenhagen (2009) was internationally cited as having limited success, its 'Accord' committed to establishing a Green Climate Fund to 'support projects, programmes, policies and other activities in developing countries related to mitigation including REDD-plus,[3] adaptation, capacity-building, technology development and transfer'. This paved the way for the establishment of a Technology Mechanism, guided by a country-driven approach, and based on technological co-operation and national priorities. The decisions under the Durban Platform for Enhanced Action (2011) brought to fruition much of the progress made within the Bali Action Plan (2007), Copenhagen Accord (2009) and Cancun Agreements (2010) by committing to a full operationalization of the Technology Mechanism in 2012.

The Paris Agreement (2015) introduced an inclusive approach to technology co-operation via the recording of NDCs through a global registry. The diplomatic efforts resulted in an agreement that looks forward, via NDCs, and looks back by considering countries' historical responsibility. The NDC platform committed all parties to the same objectives where parties designed their own targets and solutions to meet climate goals, therein utilizing diplomacy to promote political solidarity. This enhanced transparency around the tracking of progress provided accountability for nationally designed targets. They included provisions of financial support on adaptation technology to enable developing countries to prepare for climate impact. The challenge around meeting the objectives agreed within the Paris Agreement raised a number of diplomatic challenges around how to share the climate stabilization challenge, how to balance priorities and how to engage domestic policymakers on what actions are needed to overcome barriers to meeting stabilization objectives.

Overcoming barriers through climate technology diplomacy

There are a number of factors that influence the proliferation of climate technologies and prevent them from competing in the marketplace and achieving the necessary large scale deployment. Such factors are attributed to the environment for technologies and often are dependent on the people, systems, techniques and knowledge and overall conditions that are required for mass uptake and deployment. While hardware deployment within developing nations can reflect lessons from developed countries' experiences, the barriers and obstacles that are encountered, and the tools to enable full transfer, differ across countries.

Climate technology deployment can flourish within a suitable 'enabling environment'[4] at a national level. The tools to enable full technology transfer, and the corresponding barriers, differ significantly across countries. These barriers often include endogenous capacity constraints, regulatory barriers, a lack of information and policy uncertainty. Technology co-operation and government collaboration can overcome many of the deficiencies. Even so, deficiencies remain in our understanding of institutional frameworks and the elaborating measures to overcome these barriers (Boldt et al. 2012).

Establishing appropriate market conditions or 'enabling environments' is a requirement to transfer climate technologies. The UNFCCC INDC Synthesis Report (2016) (on the aggregate effect of 161 (I)NDCs of the 189 parties) focussed on parties undertaking action to address climate change including renewable energy and efficiency, sustainable transport, carbon capture and storage, conservation and the sustainable management of forests. Enhanced co-operation through a multilateral response is necessary. The Synthesis Report expressed how several countries may have to overcome a range of economic, technological and capacity-related barriers to achieve the objectives of their NDCs. Many Parties indicated in their (I)NDCs the general role of cooperation related to financial, technology transfer and capacity-building support for implementing domestic measures to achieve the objectives of their NDCs. The earlier Second Synthesis Report to the UNFCCC on Technology Needs Identified By Countries (2013) categorized the predominant technology barriers across 39 countries as economic and market (high cost of technologies and limited state resources), human (inadequate capacities of personnel), and information and awareness (shortage of information on climate technologies and energy efficiency).[5] Undoubtedly, establishing an adequate institutional and regulatory framework, alongside accompanying economic incentives, can foster and stimulate the required deployment of climate technology and the associated growth.

The establishment of the UNFCCC 'Technology Mechanism', operationalized in 2012, recognized the importance of building endogenous capacities and mobilizing private sector investment to enable developing countries to deploy climate technologies. Successful technological deployment generally implies the implementation of the adopted technology resulting in increased productivity in its use and the creation of the long-term capability to manage that technology (Foray 2009). These capabilities strengthen environments that stimulate the transfer and acceleration of climate technology in developing countries that are currently in the process of developing such frameworks as part of their Technology Action Plans. Establishing a network that can enable capacities to be built upon, best practices to be deployed, and the skillset to replicate and scale pilot projects was core to the establishment of the UN Climate Technology Centre and Network (CTCN), discussed below.

Accelerating the diffusion of, and further innovation in, climate technologies across all sectors requires enhanced diplomatic and cooperative action. Such action enables the mobilization of private sector investments for deploying technologies and innovation to accelerate the market take-up of better climate technologies and breakthrough solutions in countries. There is global consensus that both public and private investments that support the deployment of, and innovation in, carbon intensive technologies must be shifted towards climate technologies. Such co-operation requires open dialogue and pragmatic linkages between the Technology Mechanism and the UNFCCC's Financial Mechanism; and, at the operational level, between the CTCN and the Green Climate Fund and Global Environment Facility. Co-operation brings coherence across reporting requirements for climate technology and other topics (including all means of implementation and measurement, reporting and verification (MRV) of support).

Climate technology diplomacy and co-operation under the UNFCCC technology mechanism

Responding to adverse risks of climate change requires a multi-sectoral diplomatic approach with a number of strategies to be deployed. Many of the climate technologies that have, potentially, the largest impact are mature and well understood. However, the conditions in which they can be accelerated, the capacity required (within countries) to facilitate deployment, and the research and innovation to improve the identification of, and access to, such technology are more opaque and differ by jurisdiction. While scientists accept that one-third of mid-21st century warming is locked in, given the amount of greenhouse gases within in the atmosphere, the remaining two-thirds are dependent on what is discharged in the coming two decades (Sims Gallagher 2009). This brings a realization that LDCs, SIDS and emerging economies will play a significant role in the future climate action. Indeed, the resulting 'beneficiaries' of accelerated deployment of both mitigation and adaptation technologies should be LDCs and SIDS.

In 2010, as part of the Cancún Agreements, the Conference of the Parties to the UNFCCC agreed to establish a Technology Mechanism to facilitate the implementation of actions for enhancing technology development and transfer to support mitigation and adaptation activities in developing countries, including research, development, demonstration, deployment, diffusion and transfer of technology, and based on nationally determined technology needs (UNFCCC 2010).

In 2012, the UNFCCC established the Technology Mechanism to promote the acceleration and transfer of mitigation and adaptation technologies to developing countries. Its operationalization is clearly aligned with the co-operative frameworks that facilitate private and public sector investments. It focusses attention on creating 'enabling environments' or institutional contexts that facilitate the transfer of technology into developing countries. Through establishing technology co-operative mechanisms, it facilitates frameworks that enable

Governments[6] to manage technological change and so avoids 'institutional inertia', a concept coined by Perez (1983) and Freeman (1997), as a constraint on adoption of climate technologies.

The Mechanism provided for the establishment of a policy arm, the *Technology Executive Committee (TEC)*, and an operational arm aimed at climate technology diplomacy through co-operative action, the *Climate Technology Centre and Network* (CTCN). One of the CTCN's core services involves fostering collaboration among climate technology stakeholders via its network of regional and sectoral experts.

The CTCN was established to provide the supporting frame to enable Parties to stimulate the development of the national skills base and improve local access to climate technologies by providing the technical assistance to develop policy, legal frameworks and the institutional base for transfer. The appropriate mechanisms include innovative financing, endogenous support and R&D stimulation. Developing countries were provided with a facility to strengthen their technical capacity and institutional capacity for climate technology transfer, develop policy frameworks that incentivize investment and enhance private finance, and develop the infrastructure to support the diffusion of technology.

The CTCN diplomatic effort involves engagement across 158 countries via identified technology focal points within countries, titled national designated entities (NDEs). This structure facilitates a network of national, regional, sectoral and international technology centres, networks, organizations and private sector entities. In terms of responding to the needs of developing countries, requests for technical assistance to the CTCN are made through their NDEs.

The structure, hosting and delivery of the CTCN is framed around climate technology diplomacy and is systemic in nature, in that it depends on linkages between different actors in a system, as well as a large number of contextual factors. The CTCN is hosted by the United Nations Environmental Programme, in collaboration with the United Nations Industrial Development Organization, and is supported by 11 partner institutions with expertise in climate technologies.[7] The consortium provided initial support but the vast majority of the estimated operational costs of $100 million for the first 5 years is funded through voluntary donor contributions. In 2017, the support of $50 m is being made from the EU, 'Umbrella Group' countries[8] and Korea. Support from the Financial Mechanism is very limited.[9]

The UNFCCC TEC advises that activities related to the technology cycle, policy and regulatory frameworks and financing should be considered in an integrated manner and that the capacity of developing countries to assess, absorb and develop technologies needs to be enhanced. Byrne et al. (2012) emphasized the importance of the Mechanism's alignment with national initiatives and policy frameworks: 'Internationally-driven policy initiatives such as the Technology Mechanism of the UNFCCC, and its associated CTCN, open up opportunities to help build low-carbon innovation systems . . . these will need to be aligned and synergistic with national policy frameworks if developing countries are to realize self-determined low-carbon innovation.' Enhancing cooperation between developers, suppliers and consumers remains critical as countries create the predictable policies that enable greater market and system integration and higher penetrations of climate technologies.

Many developing countries may lack the technological capability to immediately stimulate and deploy such technologies, even with the appropriate co-operative mechanisms, policy and economic incentives in place. Governments can improve the cost effectiveness of options through framing technological choices and implementing behavioural change programmes (Anderson 2006). While government intervention tends to focus on national interests, it is often non-economic barriers that impact on the success of policy incentives irrespective of their design, and this draws attention to the importance of the country-specific contexts. Ultimately, there is a requirement to understand the nature of public attitudes and beliefs with regard to behavioural change. The establishment of the UNFCCC Technology Mechanism, notably the CTCN, through climate technology diplomacy target many of these barriers and impediments in the guise of the provision of technical assistance, capacity building activities and knowledge management.

Beyond the Paris Agreement

The Paris Agreement established a new Technology Framework to provide overarching, strategic guidance to the work of the Technology Mechanism under the Paris Agreement and to help to present a vision that is ill defined within the Paris Agreement. This new Framework builds on an existing one, established at COP 7 as part of the Marrakesh Accords. The new Framework was developed to guide implementation the transformational changes envisioned in the Paris Agreement. There remains a little confusion about its coexistence with other provisions under the Convention or the Paris Agreement that could potentially frame a more narrow understanding of the Framework, potentially resulting in a Standing Committee of Technology to enhance coordination inside the Technology Mechanism. The Framework needs to be flexible to respond to changes over time but also coherent, transparent and transformative in terms of its principles.

The Technology Framework presents an opportunity to integrate the different means of implementation and the cross-cutting role of climate technology diplomacy to reach the mitigation and adaptation objectives of the Paris Agreement. This presents co-operative potential in terms of enhancing the coherence of the Technology Mechanism with ongoing work both under and outside the Convention. The intent is to improve the effectiveness and efficiency of technology development and transfer under the UNFCCC and its interaction with other UN bodies, parties and the private sector. This would help to address the existing fragmentation between the Technology Mechanism, the processes around Technology Needs Assessments, Technology Action Plans and the implementation of technology components in Nationally Determined Contributions. The Framework can further guide implementation of the transformational changes envisioned in the Paris Agreement and help to reflect exemplar use cases of global cooperation and international innovation systems. Parties will elaborate the framework further at future negotiations with the aim to finalize it at COP24 in December 2018.

Climate Technology Diplomacy beyond the UNFCCC

The IEA's Technology Collaboration Programme

For over 42 years the International Energy Agency (IEA) has led the acceleration of climate technology through multilateral co-operation and diplomatic effort in the form of energy partnerships and initiatives. Such effort aims to accelerate energy technology innovation, support economic growth and enhance energy security. The IEA through its Technology Collaboration Programmes (TCPs)[10] engages in collaborative effort with the public and private sector to share knowledge, work collaboratively and pool resources to deliver integrated, cost effective solutions to common challenges. The decarbonization imperative requires a global transition and international diplomatic effort. Such effort takes the form of strong, resource-efficient, and results-orientated multilateral co-operation.

Targeted climate technology actions include energy efficiency, renewable energy, fossil fuels, fusion power and cross-cutting issues. Deliverables included the development of inventions, proof-of-principle; pilot and demonstration projects, field studies, measurements and instrumentation; and applied research and standards development, disseminated via scientific publications, guidebooks, databases and policy papers.

The IEA network of countries expanded this flexible collaboration mechanism and widened it to include both member and partner countries and international industry. The IEA has close to 40 TCPs operating that involve climate technology collaborative actions, involving about 6,000 experts from government, industry and research organizations in 51 countries. The TCPs have examined more than 1,900 energy-related topics and engaged sponsors including industry associations, engineering and service companies, multinational enterprises, privately funded research organizations and consultancies.

International Renewable Energy Agency's Regional Action

The International Renewable Energy Agency (IRENA) serves as the principal platform for international co-operation through a sharing of information concerning member renewable energy policy, technology and resources. Its primary role in climate technology diplomacy involves the facilitation of dialogue, knowledge sharing, enhancing capacity and sharing of best practices at the global, regional and national levels. The principles of cooperation and partnership are essential underpinnings of IRENA's work. This co-operation is delivered at a regional level where IRENA achieves economies of scale in the deployment of renewable energy technologies and adopts an integrated approach to issues such as energy trade, regulatory frameworks and policies, and regional power infrastructure markets.

IRENA's 152 country members collaborate with stakeholders and development partners to enhance synergies, facilitate convergence of effort, catalyse action on its regional approach to the deployment of renewables, and strengthen their effectiveness through strategic partnerships. Within its work programme for 2016–2017,[11] IRENA has committed to continue its contribution to global and regional initiatives, such as the Clean Energy Ministerial (CEM), G7 and G20 processes and The World Future Energy Summit, the Africa Renewable Energy Initiative (AREI).

Mission Innovation challenges

Mission Innovation is a global partnership, launched during the 2015 Paris Climate Change Conference, aimed at reinvigorating and expediting the development of clean energy technologies to make them more affordable while creating commercial opportunities. Mission innovation was established in November 2015 to catalyse global research efforts that could provide significant benefits in reducing greenhouse gas emissions, increasing energy security and creating new opportunities for clean economic growth. The focus of the 22 member countries (and the European Union) is on strengthening collaboration and driving engagement from the research community, industry and investors and seeking a doubling in their governmental and/or state-directed clean energy research and development investment over 5 years. Mission Innovation's challenges are intended to support, reinforce and compliment national-led activities through encouraging mutually beneficial engagement across areas such as building and technology innovation roadmaps, joint research and public–private partnerships and joint member research. Challenges focus on smart grids, off grid access to electricity, carbon capture, sustainable biofuels, clean energy materials and affordable heating and cooling.

Climate technology diplomacy is central to delivering on the Mission Innovation objectives and challenges as national governments cooperate with each other and with international organizations, businesses and the energy industry to drive progress in the global clean energy transition. A number of campaigns have been announced in co-operation with the Clean Energy Ministerial including the Global Lighting Challenge, Energy Management Campaign, Corporate Sourcing Renewables Challenge and the Net Zero Emission Buildings Commitment Campaign.

At its June 2017 Ministerial meeting, Mission Innovation member countries agreed to collaborate with the World Economic Forum, an international organization for public–private cooperation, to engage private sector members from the Forum's Communities with Mission Innovation. Details of the collaborative action included collaboration on a joint high-level platform for strategic public–private dialogue and a facilitation of a sharing of knowledge research, information and technological development on clean energy.

Announced concurrently with Mission Innovation was the Breakthrough Energy Coalition, a group of 28 high net worth individuals from 10 countries who have pledged billions in private capital investments for clean energy. Led by philanthropist Bill Gates, the coalition will focus on early-stage technologies for transportation; industry; agriculture; and power generation, storage and system efficiency in Mission Innovation countries.

Conclusion

The world is at a political tipping point in terms of meeting our future climate stabilization challenges. There is political acceptance of the scale and urgency of the climate action required. Significant diplomatic efforts are required given the number of deficiencies that exist in terms of meeting the stabilization objectives identified by Parties within their NDCs. There seems to be insufficient ambition reflected in NDCs that will not result in meeting the global stabilization trajectory. Taken as a collective, the NDCs are insufficient in terms of the variability of clarity of intent as some remain sector specific and others are country wide. The financial contributions are vague in terms of the diplomatic effort and public finance required to be mobilized. Finally, there are no penalties or negative incentives for the failure to implement NDCs.

We are at a regulatory tipping point also. The energy transition involves both the pursuit of the regulation of greenhouse gases as a pollutant, while also the selection of future pathways resulting in net zero emissions and decarbonization. Both leadership and collaborative partnerships are required to overcome any deficiencies. This ultimately requires multilateral diplomacy to ensure robustness. Undoubtedly, bilateral diplomacy can also stimulate trade and accelerate the development of technology transfer. We are at a tipping point in terms of technological availability and market supply of low carbon technologies to meet the required climate drivers.

The 5-year Global Stocktake, announced under the Paris Agreement, will determine the progression in implementation of the agreed commitments. The new Technology Framework will provide overarching guidance to the work of the Technology Mechanism in promoting and facilitating enhanced action under the Paris Agreement. Climate technology diplomacy can help shape the design of domestic policies and help frame the actions required to meet Parties' objectives. Such actions range from renewable energy deployment, energy efficiency standards, actions and conservation, smart agriculture, sustainable transport and the efficient deployment of carbon pricing and emissions trading. A significant diplomatic effort involves actions to optimize the grid and a replacement of existing infrastructure while providing safe and reliable energy. While public opinion is changing and there is an encouraging belief in the need for transition in our energy system, this needs to be strengthened with a renewed commitment by parties to enable pathways to be achieved within the urgent timeframe.

Notes

1 Article 4, Paragraph 12, Decision 1/CP.21 Paris Agreement.
2 Available mainstream technologies, such as cooking stoves, biomass devices and solar lanterns, could be refined in the near term and adapted to meet the needs of developing countries.
3 REDD-Plus is the United Nations Collaborative Programme on Reducing Emissions from Deforestation and Forest Degradation in Developing Countries. REDD-plus includes Reducing Emissions from Deforestation and Forest Degradation in Developing Countries; and the role of Conservation, Sustainable Management of Forests and Enhancement of Forest Carbon Stocks.
4 An 'enabling environment' is defined as a conducive institutional context that embodies the appropriate institutional, policy, market and regulatory conditions for the uptake and deployment of low carbon technologies. The UNFCCC (Article 4.5 and 4/CP.7) defines enabling environments as: 'fair trade policies, removal of technical, legal and administrative barriers to technology transfer, sound economic policy, regulatory frameworks and transparency, all of which create an environment conducive to private and public sector technology transfer'.
5 Under the UNDP Technology Needs Assessments Synthesis Report [2010], 83% of developing countries identified economic and market conditions as the key barrier to the transfer of LCT. A further 74% of countries identified a lack of information and awareness as key barriers.
6 A successful deployment of climate technology requires governments to influence and shape the people, systems, techniques and knowledge and overall policy conditions. Such action could involve the provision of signals through stable energy investment policies, and the development of the national skills capacity by developing policy, legal frameworks and the institutional base for transfer. Governments have a significant role that involves the provision of signals through stable energy and environmental policies that stimulate low carbon investment and create new product markets.
7 Asian Institute of Technology (AIT) – Thailand/Bariloche Foundation (BF) – Argentina/Council for Scientific and Industrial Research (CSIR) – South Africa/The Energy and Resources Institute (TERI) – India/Environment and Development Action in the Third World (ENDA-TM) – Senegal/Tropical Agricultural Research and Higher Education Center (CATIE) – Costa Rica/World Agroforestry Centre (ICRAF) – Kenya/Deutsche Gesellschaft für Internationale Zusammenarbeit (GIZ) – Germany/ Energy Research Centre of the Netherlands (ECN) – The Netherlands/National Renewable Energy Laboratory (NREL) – United States of America/UNEP Risø Centre, including expertise from UNEP-DHI Centre (URC) – Denmark.
8 http://unfccc.int/parties_and_observers/parties/negotiating_groups/items/2714.php.
9 To review the first 4 years of operation and the hosting through UN Environment, a review task has been tendered to Ernst &Young. A final report will be available in late 2017.
10 Formerly called IEA Implementing Agreements (IAs). For more information see www.iea.org/publications/freepublications/publication/TechnologyCollaborationProgrammes.pdf.
11 For more information on the IRENA work programme see www.irena.org/DocumentDownloads/A_6_4_Work_Programme_and_Budget_2016_2017.pdf.

References

Boldt, J., Nygaard, I., Hansen, U.E. and Traerup, S. (2012). *Overcoming Barriers to the Transfer and Diffusion of Climate Technologies*. TNA Guidebook Series, UNEP Riso Centre.

Byrne, R., Schoots, K., Watson, J., Ockwell, D., de Coninck, H., Sims Gallagher, K. and Sagar, A. (2012). Innovation systems in developing countries. Climate Strategies Policy Brief as part of Climate Technology and Development project. Available via www.climatestrategies.org.

Carpenter, C. (2012). *Taking Stock of Durban: Review of Key Outcomes and the Road Ahead.* UNDP Environment and Energy Group, April 2012.

De Coninck, H., Fischer, F., Newell, R. and Ueno, T. (2008). "International Technology-Oriented Agreements to Address Climate Change." *Energy Policy* 36, 335–356.

Foray, D. (2009). *Technology Transfer in the Trips Age: The Need for New Types of Partnerships Between the Least Developed and Most Advanced Economies.* ICTSD Programme on IPRs and Sustainable Development.

Freeman, C. (1997). *The Political Economy of the Long-Wave*, in: A. Tylecote and J. van der Straaten (eds), *Environment, Technology and Economic Growth.* Edward Elgar, Cheltenham, UK, pp. 159–183.

Haites, E., Duan, M. and Seres, S. (2006). "Technology Transfer By CDM Projects." *Climate Policy* 6, 3, 327–344.

Nicholls, M. (February 2008). "Two Years and Counting." *Environmental Finance.*

OECD. (2015). *Environment at a Glance 2015.* OECD Indicators.

Perez, C. (1983) "Structural Change and Assimilation of new technologies in the economic and Social System." *Futures*, 15, No. 4, October, 357–375. In Italian, in: P. Bisogno (ed.), *Paradigmitecnologici: Saggisulleconomiadel Progresso Tecnico.* Prometheus No. 2, 1985, Milan, Italy, pp. 155–186.

UNFCCC. (2007). "Bali Action Plan." Conference of the Parties to the UNFCCC.

UNFCCC. (2010). "Cancún Agreements, decision 1/CP16." Conference of the Parties to the UNFCCC.

UNFCCC. (2013). "Second Synthesis Report to the UNFCCC on Technology Needs Identified By Countries." Background Note on Technology Briefs. Fccc/Sbsta/2009/Inf.1.

UNFCCC. (2016). "INDC Synthesis Report." http://unfccc.int/resource/docs/2016/cop22/eng/02.pdf.

World Business Council for Sustainable Development. (2010). *Enabling Frameworks for Technology Diffusion: A Business Perspective.* WBCSD, Geneva, Switzerland.

10 Technology and the Paris Agreement

From means of implementation to climate innovation to transformation

Karsten Krause

Introduction

In order to avoid a 'dangerous interference with the climate system', the international community has agreed to limit the global mean temperature increase since pre-industrial times to well below 2°C. Better technologies and new ways of using existing knowledge are an important means to cut greenhouse gas emissions and enhance climate resilience. New or better technologies and innovative ways of accelerating their dissemination have an essential role to play in meeting climate objectives, as well as in creating new jobs and sustainable growth.

As a pillar of the United Nations Framework Convention on Climate Change (UNFCCC), 'Technology' has been established to facilitate North–South, South–South and triangular cooperation. The UNFCCC negotiations, in particular the Paris Agreement's chapter on technology transfer and development, create a broad framework for greater global collaboration. Since the beginning of the climate negotiations in 1992, the critical role technology must play is recognized. Article 4.5 of the UNFCCC requires developed countries to 'take all practicable steps to promote, facilitate and finance, as appropriate, the transfer of, or access to environmentally sound technologies and know-how to other Parties, particularly developing country parties to enable them to implement the provisions of the Convention.' At the UN climate conference in Cancún in 2010, Parties to the UNFCCC agreed to implement a Technology Mechanism to facilitate enhanced action on technology development and transfer. The Technology Mechanism is made up of two components: a Technical Executive Committee, comprising 20 elected experts; and a Climate Technology Centre and Network (CTCN) which helps developing countries access information on mitigation and adaptation technologies. Requests for technical assistance range from designing renewable energy policies and biodiversity monitoring to saving mangrove forests for coastal protection.

The Paris Agreement includes an explicit recognition of technology development and transfer in Article 10. It commits nations to 'strengthen cooperative action' on climate technology. It introduces both a 'long-term vision' on technology's contribution to climate resilience and mitigation and a 'technology framework' to guide the work of the Climate Convention's existing Technology Mechanism. The Agreement puts a new focus on innovation as 'critical for an effective, long-term global response to climate change'. It commits the Climate Convention's technology and financial bodies to support R&D and developing countries' access to technology, 'in particular for early stages of the technology cycle'.

Under the Paris Agreement, how the mobilization of finance and the provision of resources to developing country Parties (Art. 9) can be handled is separate from the enhancement of capacity in developing countries (Art. 11) and Technology Development and Transfer (Art. 10). While the importance of technology development and transfer is recognized, only an integrated perspective on the possible contribution to the Paris Agreement's objectives, the interdependency within the means of implementation under the Convention and global innovation systems can help to enhance understanding.

Supporting decoupling and transformation

While some international environmental conventions are able to narrow the problems and solutions to a defined list of technologies, climate technology remains a fuzzy concept. Climate technologies are a broad cluster of products and capacities that include both mitigation and adaptation technologies, or technologies that contribute to both. They form an integral part of the fight against climate change and are a key driver for sustainable growth. The use of climate technology results in lower greenhouse gas emissions and a higher resilience to climate change than business as usual. The incremental contribution to climate objectives is often intertwined with a prime utility. Wind turbines produce electricity, but with less greenhouse gas emissions compared to business-as-usual, i.e. a gas-fired power plant. The incremental contribution to climate target and the additionality to baseline scenario are the key characteristics of climate action.

While energy-related technologies, such as LED lighting systems, insulation for buildings or solar panels are obvious examples, the Paris Agreement's objective implies that all sectors need to contribute to the greenhouse gas emission reduction efforts. This includes industry, transport and agriculture. The resulting wide array of mitigation options are complemented by technologies for adaptation to the adverse impacts of climate change. These could address resilience against the changing climate conditions or against extreme weather events. Examples are effective irrigation systems, salt water-resistant crop varieties or disaster response measures. In this broad interpretation of climate technology, the dividing line 'cannot be clearly defined'.

'Transformation' has become a buzzword in debates about climate action and sustainable development. But while the term has become very popular, it is often unclear what is meant exactly. The Paris Agreement includes several references to transformational change. 'It refers to a balance between anthropogenic emissions by sources and removals by sinks of greenhouse gases in the second half of this century' (Article 4). But it also includes the aim of 'making finance flows consistent with a pathway towards low greenhouse gas emissions and climate-resilient development' (Article 2.1(c)). Beyond its role as part of the Agreement's means of implementation, Article 10 on 'Technology Development and Transfer' addresses the contribution of technology in a dynamic perspective, for example in Art. 10.5: 'Accelerating, encouraging and enabling innovation is critical for an effective, long-term global response to climate change and promoting economic growth and sustainable development'.

In this context, 'Technology Development and Transfer' looks in parallel to decoupling the currently available solutions from emissions-intensive resource use and a transition towards alternative systems to avoid 'dangerous interference with the climate system'. The expected contribution to the objectives of the Paris Agreement is not only a better use of resources for continuing a business as usual ('relative decoupling'). There is also a need for 'absolute decoupling'. Different products, services and lifestyles are needed to ensure sustainable supply for the growing world population. The provisions in Article 10 aim also at the long-term perspective through finding different technological and organizational solutions. Given the ambition level of the Paris Agreement, fundamentally different approaches are needed to respond to the needs of the growing world population. Global climate policy and the transfer into national programmes can give short- and long-term guidance. In particular investments in infrastructure, such as airports, roads or energy networks manifest a development pathway. The development, testing and acceptance of low emission alternatives are long-term processes and embedded in broader economic development pattern and enabling environments at the place of technology deployment.

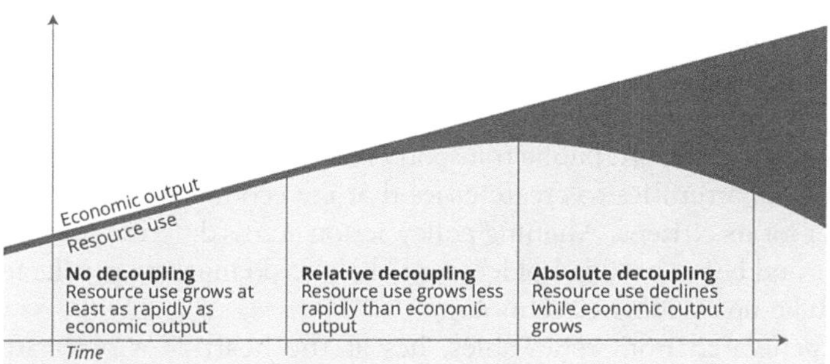

Figure 10.1 Relative and absolute decoupling

Source: European Environment Agency (2015) SOER 2014 – The European environment – state and outlook 2015.

Technological change can help to improve the efficiency of resource use and substitute a polluting activity with a non-polluting activity, and notable examples have included the removal of tetraethyl lead from automotive fuel and CFCs from refrigerants. It is also possible to envisage a scenario in which economic growth is decoupled from the use of fossil fuels and related CO_2 emissions by switching to 100% renewable energy, but this is not the same as decoupling GDP growth from energy use. While air transport needs to become more efficient in order to reduce the carbon footprint of a business trip, a video conference can replace the trip. In particular, digitalization can help to 'dematerialize' needs. But the hidden environmental footprint of producing IT equipment and the energy consumption needed to provide internet services may offset potential mitigation benefits. Therefore, a decoupling requires a transition to post carbon production and consumption models in a systemic approach.

Key technology clusters for transformational action

The next 15 years of investment will determine the possibilities to change or to locking in the stock of high-carbon assets. Delay in reducing emissions makes it progressively more expensive to shift towards a low-carbon and climate change resilient economy. However, while emissions are falling in Europe and other regions, it is predicted that by 2020, nearly two-thirds of the world's emissions will come from emerging and developing countries, where the majority of the infrastructure of 2050 has not been built or planned. It is therefore vital, to foster decoupling not only through an improved efficiency but also through products and services that bring a deeper change.

While there is no consensual list of mitigation and adaptation technologies, the Technology Executive Committee, the UNFCCC's technical expert meetings[1] and the Global Climate Change Action Agenda[2] foster thematic action complementary to the negotiations. Beyond individual technologies, three technology systems are often considered for their huge potential for structural transformation and technological change: cities, electricity and agriculture:

- Cities generate around 80% of global economic output, and around 70% of global energy use and energy-related GHG emissions (New Climate Economy Report 2014). Cities are also vulnerable to extreme weather events. But smarter individual transport, public transport systems and infrastructure over technological opportunities to create cities that are economically dynamic and healthier for its citizens. Aligning policy action across different levels of governments and between stakeholders could help to do much to accelerate the market take-up of climate technology.
- Electricity, generated from renewables, lies at the heart of a successful decarbonization of energy systems. But often electricity markets do not deliver the long-term price signal needed for de-risking the investment in high capital-cost, low-carbon technologies. Ensuring competitive and timely investment in low-carbon solutions requires suitable market arrangements

such as long-term supply agreements and a robust CO_2 price signal. Rapidly falling costs, particularly of wind and solar power, could lead renewable energy sources to account for the majority of all new electricity generation.
- Agriculture and land use productivity will determine whether the world can feed a growing world population while sustaining natural environments. Food production can be increased, forests protected and land use emissions cut by raising crop and livestock productivity, using new technologies. Sustainable land-management practices – reduced deforestation, restoring degraded land, low-carbon agricultural practices and increased carbon sequestration in soils and forests – can make a large contribution to reducing greenhouse gas emissions while responding to growing food demands. They could also improve the resilience of our economies to a changing climate by protecting ecosystems.

Emission reduction potential can be identified across all sectors. Cross-cutting to other sectors, transport and information/communication technology provide enabling services to citizens and companies. At the same time, both sectors are among the fastest growing source of emissions. If both sectors would be a country, they would be already among the top global emitters.

Supporting innovation

Innovation has been recognized as a process, combining different factors to influence the applicability and market absorption of a technology. In this process, a technology passes different readiness levels, from a research and development phase, over demonstration towards deployment, and diffusion of a technology. Innovation involves multiple actors interacting in various ways and with multiple feedback loops between and across the various stages (TEC 2017). It combines public investments in the education system, financial support for research and innovation projects, enabling policy frameworks and incentives for entrepreneurial activities in evolving combinations, combining experiences and experimental interactions. From early stages of development towards project refinancing, the perceived risks are changing from the capability of a solution towards commercial risks. In particular, the high up-front investments required in many forms of renewable energy compared to the rather low running costs during the operational phase make appropriate risk management key for innovation. At early stages or within the 'Valley of Death', support is dominated by public sector contributions in the form of grants. The different support mechanisms aim to 'crowd in' other public actors, in particular from the private sector. Many grant-based support schemes require co-financing. In particular financial instruments, such as loans and guarantees, aim to leverage private investments. In accordance with the declining risks, the ratio between public and private engagement is changing. The role of the 'state' is changing from being a research funder towards the manager of the investment environment, enabling the market take-up of climate technologies. This enabling environment secures investment and sets incentives to stimulate innovation. For some climate technologies, the incremental contribution to climate objectives can be rewarded by an emissions trading system, like in the European Union. Others, like energy efficiency might be pushed better through minimum standards. Indirect incentives can also lead to investments in climate technologies. For instance, a higher water price would push the sales of smart irrigation systems. For most technologies, direct co-benefits, like energy savings, are additional motivations. Prevented air pollution and job creation are complementary benefits of ambitious climate policies.

National characteristics have a significant influence on shaping innovation outcomes. National innovation activities depend on size and income level. Smaller countries or countries with limited national research investments may be more focused on incremental and adaptive innovation, while richer countries can support early stages of technological development and can easier provide a critical mass to push the technological frontier (TEC 2015). The non-linear interaction of cultural, organizational, behavioural and technological factors can be considered as a National System of Innovation or as the enabling environment of activities along the technology cycle. The UNFCCC understanding of technology always includes software and organizational aspects. Education, from science classes to industrial doctorate programmes, is important to utilize and co-develop a technology. In particular, engineering education and vocational training are important as bases for innovation. Key factors are, apart from the financial means available, the design and cooperation among national actors, e.g. the flow of information and knowledge, and collaboration between firms, universities and research institutes. Comparisons between countries and regions show that different national innovation-related polices, education investments and infrastructures

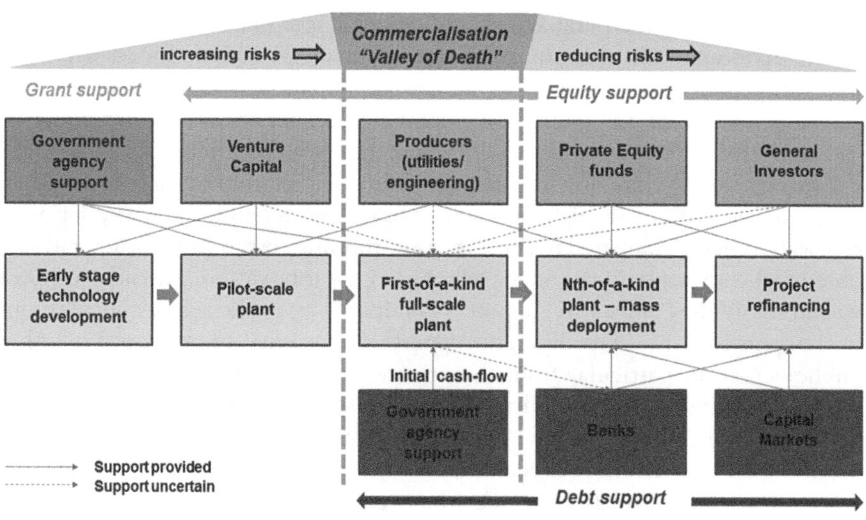

Figure 10.2 From research laboratories to mass deployment

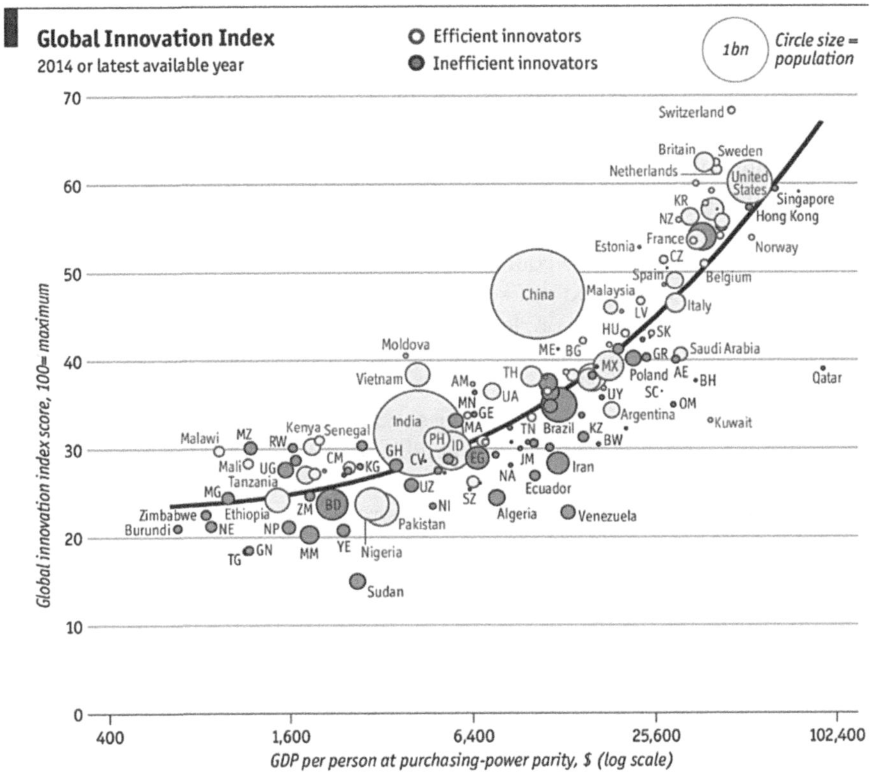

Figure 10.3 Comparing countries according to the Global Innovation Index

can help one country to perform on par or better than some neighbouring countries or their development-level peers from other continents (GII 2017).

The effects of public funding can be enhanced through co-ordination and improved public governance at all levels of government. Early engagement by key stakeholders in goal setting and planning, at every level of government, will help ensure the relevance and consistency of policy objectives and expectations in markets. Enhanced co-ordination and improved public governance, especially among environment and natural resource management, energy and investment authorities, are also critical factor for low-carbon investments. This will be particularly important at the metropolitan level to ensure a good integration of transport and spatial planning policies for low-carbon cities, or in the context of solving trade-offs in rural land use.

Policy support to climate-friendly practices can leverage companies' contributions to addressing climate change over and above the influence of regulatory approaches such as carbon pricing. For instance, corporate carbon disclosure is an important element of responsible business, and should be increasingly encouraged by governments and investors.

Creating enabling environments

Enabling environments are location-specific. An established solution in one country could be unknown, not available or not accepted in another. The interaction of entrepreneurs, universities, schools, governments and technology users creates an ecosystem. A stable regulatory framework is a precondition for transferring intellectual property and investing in a country. Many climate technologies show a particular vulnerability to instability, given the high initial investment costs. For many countries, the 'import' and adaptation of a solution can lead to an accelerated market introduction and take-up of climate technologies. Focus should be on learning and first-to-market approaches, where investment models, risk management an application models reflect national circumstances and capabilities.

Information about climate technologies and policy measures are available. Under the UNFCCC, solutions are discussed at different places. Already since 2001, more than 80 developing countries have conducted Technology Needs Assessments (TNAs) to identify national technology strategies to address climate change challenges. TNA aim to determine the mitigation and adaptation technology priorities of a country. TNAs help to track needs for new equipment, techniques, services, capacities and skills necessary to mitigate GHG emissions and reduce the vulnerability to climate change. TNAs identify a portfolio of projects and programmes to support the transfer of technologies, usually prepare by consultants for the national government in consultation with stakeholders. TNAs and resulting action plans form also important inputs for other national processes related to the UNFCCC: nationally appropriate mitigation actions (NAMAs), national adaptation plans (NAPs), and national communications (NCs), and to develop National Determined Contributions (NDCs). While over 160 countries submitted first NDCs, only 80 countries conducted a TNA. In comparison to NDCs, TNAs are developed on a well-established methodology and focus more on pathways and less on establishing targets (TEC 2016). There is a need to engage the financial and business communities and funding sources under and outside the Convention to facilitate the implementation of the TNA outcomes.

A second component to coordinate national actors to enhance the enabling environment are the National Designated Entities for climate technology development and transfer.[3] NDEs, nominated by over 150 UNFCCC Parties, are linked to the CTCN. They play a fundamental role as access point and coordination hub for technology. The nominated individuals ensure that requests for technical assistance submitted to the CTCN reflect their national circumstances and priorities. NDEs also ensure that support provided by the CTCN is well coordinated at the national level with other processes that address climate change (CTCN 2017). A key task is to ensure the engagement of relevant ministries, focal points for other UNFCCC mechanisms, the private sector, civil society and academia as appropriate in a request. This function is organized differently by different Parties. While many Parties locate the NDE in an environmental or climate ministry, other countries see more the technology aspect and locate the NDE in economic affairs departments or a specialized agency.

Through the NDE, the CTCN helps developing countries to prioritize their technology needs, research potential regulatory and policy barriers and support technology deployment (including identifying sources of additional public or private funding). The Centre is able to provide these services by mobilising expertise on policy and technology support from a global network of finance, NGO, private sector and research stakeholders. How it works: local and national stakeholders in developing countries convey their clean technology-related needs to the CTCN via a national focal point. Upon receipt of a request, the Centre conducts an open bidding process among its network members and contracts them directly for their services. Network members represent a broad array of expertise in technology development, deployment, capacity building, finance, investment and policy. As the CTCN helps countries to create environments that are conducive to attracting and supporting technology implementation, many of its technology projects focus on both soft and hard technologies and take the form of training, assessments, feasibility studies and guidance on policy and regulatory structures.

In Uganda, for example, although geothermal energy prospects have been known about for more than 30 years, exploratory drilling for geothermal resources has yet to take place anywhere in the country. The government and development partners conducted surface studies, identified at least four sites with medium to high temperature hydrothermal systems and issued geothermal exploration licenses to private sector prospectors. However, this resulted in little or no exploratory drilling over the years. At the Ugandan government's request, the CTCN called upon its network member Carbon Counts, an energy and climate change consultancy, to identify barriers to geothermal development and propose solutions. Carbon Counts pinpointed a lack of a clear geothermal energy policy and legal framework and worked with stakeholders to develop a draft geothermal energy policy for Uganda, as well as a geothermal energy law and guidance on necessary institutional structures. Once the policy and legal framework has been implemented, the development of 100 MW of geothermal energy production could provide a clean, secure, reliable and affordable source of power for over 500,000 Ugandan households.

CTCN technical assistance needs to be seen as seed-funding to enable larger scale projects at a later stage, either by identifying technology partners or financial institutions to upscale and implement the results of technical assistance requests.

Private sector engagement and research cooperation

Private sector involvement is also a key driver of technological innovation and technology transfer. Even inside the EU the vast majority of innovation spending is realized by the private sector and technologies are owned and operated by private actors. A differentiated approach is needed to work with technology producers, banks, engineering companies and SMEs. Despite the business potential, a knowledge gap needs to be bridged to balance perceived risks and to leverage much higher investments than can be obtained through the public sector.

To maximize opportunities, both in the EU and in developing countries, cooperation and networking are key. Joint projects with international partner countries and exchanges of knowledge help to create synergies and allow the sharing of experiences on how to best innovate. Technology transfer is an important pillar of the UNFCCC and the Paris Agreement. In order to accelerate matchmaking opportunities between technology providers and interested parties in developing countries, the CTCN was launched in 2014 with support from the EU. Three years later, almost 200 clean technology transfers are now underway in more than 70 countries, in sectors ranging from agriculture and energy to industry and waste management. The CTCN NDEs are an access point into the national innovation system. TNAs and requests for technical assistance submitted to the CTCN highlight investment opportunities.

But climate technology transfer needs to be mainstreamed into existing national and global innovation systems. Given the incremental contribution of technology to climate targets, it would be impossible to differentiate the external benefits but a positive discrimination is necessary. Climate mainstreaming could help to incorporate climate change considerations into established or on-going programs, policies or management strategies, rather than developing adaptation and mitigation initiatives separately. The EU has, for example, agreed that at least 20% of its budget for 2014–2020 – as much as €180 billion – should be spent on climate change-related action. To achieve this increase, mitigation and adaptation actions are being integrated into all major EU spending programmes, in particular cohesion policy, regional development, energy, transport, research and innovation and the Common Agricultural Policy. The EU's development policy will also contribute to achieving the 20% overall commitment, with an estimated €1.7 billion in 2014–2015 and €14 billion over the years 2014–2020 for climate spending in developing countries (EU 2017). The progress towards reaching the 20% objective is monitored on an annual basis by measuring the performance of the EU programmes in the framework of the budgetary procedures. Member States have an obligation to report on climate spending under the five European Structural and Investment Funds according to an established methodology. For the research and innovation programme Horizon 2020, the climate mainstreaming target has been set to 35%. Horizon 2020 sees international cooperation as another important cross-cutting priority and the integration of scientists into EU projects is encouraged.

Guidance and coherence

The Paris Agreement established a new technology framework to provide overarching, strategic guidance to the work of the Technology Mechanism. In their negotiations, Parties underlined the coherent role to align technology-related activities with the long-term vision and other provisions under the Paris Agreement. This inclusive nature links it with national plans, related UNFCCC work but as well work beyond the international climate regime. Furthermore, it specifies five initial key themes: (1) Innovation, (2) Implementation, (3) Enabling environments and capacity building, (4) Collaboration and stakeholder engagement and, finally, (5) Support. All key themes are summarising technology-related aspects relevant for the CTCN and the TEC. There is no clear boarder as topics like innovation and support have different facets for each technology and over different stages of the technology cycle. Along the key themes, a better coherence of the work of the two arms of the Technology Mechanism and their mainstreaming with the wider context can achieved.

The Technology Framework is an important step towards integrating between the different means of implementation and the cross-cutting role of technology development and transfer to reach the mitigation and adaptation objectives of the Paris Agreement. Of a particular importance is the orientation towards contributions to delivering transformational changes. While the interrelation and arrangement of the framework's different component are still evolving, the need for an dynamic approach with differentiate view on different stages of the technology cycle has been highlighted by the European Union and other Parties. The strategic guidance provided by the framework can help to improve the effectiveness and efficiency of technology development and transfer under the UNFCCC. The framework provides an opportunity to address the existing fragmentation in the UNFCCC technology transfer institutions, in particular between the TM, the processes around Technology Needs Assessments and the implementation technology components in INDCs. The Technology Cycle[4] can capture the different push and pull incentives needed during its different phases of development and market take-up (European Union 2016):

- Cooperative action to promote climate technologies at early stages of technology cycle (research and development).
- This phase addresses tertiary education and research cooperation. Applied research and learning from working with climate technologies enhances national capabilities and widens the global availability of products and services.
- Demonstration and the market take-up of technology (demonstration, deployment).
- This phase addresses piloting and testing applications of climate technologies. At this stage the operational costs and institutional factors shall be addressed to prepare the market penetration of products and services.
- Up-scaling and enabling transformational changes through climate technology (diffusion and transfer).
- This phase looks into the conditions for deployment and dissemination of climate technologies. At this phase institutional factors shall allow public and private actors to access services and products (both hard and soft technologies) which have a significant contribution to mitigate greenhouse gas emissions and/or to enhance the resilience to climate change.

The gradual transition towards market take-up changes over time. It is characterized in early phases by relatively high current costs and a very small

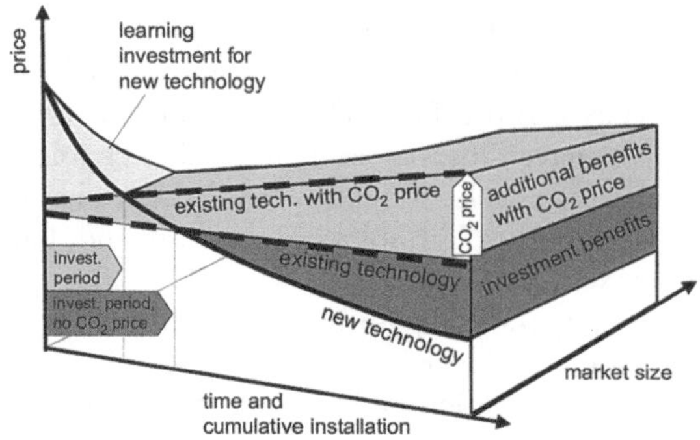

Figure 10.4 Climate technology investments through embedded learning

Source: Intergovernmental Panel on Climate Change (2007). Climate change 2007: mitigation of climate change. Contribution of Working Group III to the Fourth Assessment Report of the Intergovernmental Panel on Climate Change.

market share, and requires a high unit rate for 'learning investment'. With increasing scale and learning, costs move towards existing, higher-carbon sources, the costs of which may also be declining, but more slowly. Rising carbon prices over time bring forward the time when the new technology may be competitive without additional support, and may greatly magnify the economic returns from the initial learning investment.

In addition to a CO_2 price, the internalization of external benefits like less air pollution, local jobs or energy security benefits can help to grow the market size for a technology. The Paris Agreement provides a stimulus to create such synergies.

Conclusions

Climate technology remains a mainstreaming target in a fragmented landscape. The Paris Agreement reconfirmed the role of technology development and transfer as one of the pillars of the UNFCCC negotiations. The recognition of technology as part of the Agreement's means of implementation helps to facilitate North–South, South–South and triangular cooperation. But enabling environments are needed to incentivize innovation on a national level and to prepare the ground for collaborative action. Private investment, alongside and attracted by public investment, is seen as crucial to scaling-up climate finance and closing current investment gaps. Private investors are increasingly willing to fund climate-friendly investment, yet this significant potential capital remains relatively untapped. Removing barriers and improving the enabling conditions for attracting private investment in recipient countries is essential. As climate action is a cross-cutting objective, a mainstreaming into existing innovation policy is needed to valorize the incremental contribution to mitigation and adaptation objectives.

While it is a challenge to coordinate even the UNFCCC-related platforms to present technologies and successful climate action, the coordination between platforms under and outside the Convention has not even started. But the objective of transformational change can provide a clear guidance for the Technology Mechanism and Parties. While the TEC can help to validate options, the CTCN can ensure technology dissemination through its Network and the NDEs. Their role as national integrators is key, not necessarily as builders of the national innovation system but as sign-posts and ambassadors to the evolving global innovation system around climate technologies.

Notes

1 http://unfccc.int/focus/mitigation/technical_expert_meetings/items/8179.php.
2 http://newsroom.unfccc.int/climate-action/global-climate-action-agenda.
3 http://unfccc.int/ttclear/support/national-designated-entity.html.
4 As referred to in decision 1/CP.16, para 115.

References

Climate Technology Centre and Network. (2017). National Designated Entities. www.ctc-n.org/about-ctcn/national-designated-entities.

Deconinck, H., Fischer, F., Newell, R. and Ueno, T. (2008). "International Technology-Oriented Agreements to Address Climate Change." *Energy Policy* 36, 335–356.

European Commission. (2017). Climate integrated into all EU spending areas. https://ec.europa.eu/clima/policies/budget_en.

European Environment Agency. (2015). SOER 2014 – The European environment – state and outlook 2015.

European Union. (2016). Principles and structure of the technology framework under Article 10, paragraph 4, of the Paris Agreement, EU submission to the UNFCCC.

Global Innovation Index. (2015). Global Innovation Index report. www.globalinnovationindex.org/.

ICF. (2016). Innovative financial instruments for first-of-a-kind, commercial-scale demonstration projects in the field of energy. http://ec.europa.eu/research/energy/pdf/innovative_financial_instruments_for_FOAK_in_the_field_of_Energy.pdf.

Intergovernmental Panel on Climate Change. (2007). Climate change 2007: mitigation of climate change. Contribution of Working Group III to the Fourth Assessment Report of the Intergovernmental Panel on Climate Change.

Technology Executive Committee. (2015). TEC brief strengthening national systems of innovation to enhance action on climate change. http://unfccc.int/ttclear/misc_/StaticFiles/gnwoerk_static/TEC_only/3fc5cbbf4ccb4c43930b0ce7326476a3/2d000eae9f0f4138989ce5ec30234caf.pdf.

Technology Executive Committee. (2016). Linkages between the technology needs assessment process and the nationally determined contribution process. http://unfccc.int/ttclear/misc_/StaticFiles/gnwoerk_static/TEM_TEC_meetings/6ae9ca9c5882472e94b23c522b6ded34/71375cfd955a446db561dba61fd96c2f.pdf.

Technology Executive Committee. (2017). Draft TEC brief on innovation. TEC/2017/15/13. http://unfccc.int/ttclear/misc_/StaticFiles/gnwoerk_static/tn_meetings/102833b8f1844e3f9a4a678274cb0235/cf5e968cf22648ebba68233bd36042f8.pdf.

The New Climate Economy Report. (2014). Better growth better climate – the New Climate Economy report. http://newclimateeconomy.report/2014/.

11 Climate diplomacy and the UNFCCC institutions

The case of the Adaptation Fund

Laura Hanning Scarborough

Introduction

As climate science evolves, we learn more about climate change and the inevitable effects it has on the planet. The more the climate changes, the greater is the need to adapt, and the costlier and more complex responses become. The ambition for adaptation hinges on success or failure of mitigation to limit the temperature increase.[1] While the logic is simple, forecasts prove to be more elusive. What is clear from the 5th International Panel on Climate Change (IPCC) Report is that despite a growing number of climate mitigation policies, global greenhouse gas (GHG) emissions and temperature are rising (IPCC 2014). As such, adaptation and financing of it are gaining prominence.

As a subject of international regulation, adaptation can be traced back to the 1992 text of the United Nations Framework Convention on Climate Change (UNFCCC). References to the need to adapt, cooperate and share financial burdens in that effort are made throughout the Convention.[2] Following the adoption of the Kyoto Protocol and subsequent decisions, a number of instruments, modalities and procedures have been developed to aid international adaptation action (World Resources Institute 2017). However, it was not until the Paris Agreement that adaptation was acknowledged as one of the fundamental pillars of the international effort to strengthen the global response to the threat of climate change.[3] The Agreement seeks to balance mitigation and adaptation in terms of scope of work and support.[4]

It is suggested that the ambition to limit temperature rise to 1.5 degrees will require trillions of dollars.[5] While most of the funds are expected to come from private sector, public finance will play a critical role in catalysing support. It is therefore a right moment to reflect on the institutional framework that underpins the public international adaptation finance in order to contribute to the debate on the future institutional arrangements.

Much has been written on the emerging landscape of public institutions involved in deploying climate-related finance from analysing effectiveness of funds to the calls for streamlining of finance and calls for reorganization.[6]

The present chapter focuses on governance issues of the UN Adaptation Fund (AF) with a particular focus on ability of the Fund's board to oversee the administrative apparatus, which performs daily tasks, and to perform a role of strategic steering, reflecting the interests and aims of the Kyoto Protocol Parties.

The Adaptation Fund

Established in 2001, under the Marrakesh Accords, The UN Adaptation Fund (AF) became operational in 2009 and gained an independent legal form by an act of the German Parliament in 2011. The Global Environmental Facility (GEF) was invited to host the Fund's secretariat through a 'dedicated team of officials in a functionally independent and effective manner' and the World Bank was appointed as interim trustee to 'hold in trust and manage the funds, assets and receipts that constitute the Fund'.[7] Thus the AF administrative apparatus is not an independent 'international organization' *per se* but bears the set up similar to international organizations (IO) serving governments involved with raising, channelling, and deploying international public finance. For the purposes of this chapter and to help frame the discussion in a theoretical context, the secretariat and trustee of the AF together are referred as an international organization serving as an agent of the Kyoto Protocol Parties.

The Adaptation Fund as a multilateral climate finance instrument is innovative in its governance, resource mobilization, and allocation of funds. The AF was one of the first such institutions to appoint a balanced board membership, without preferential treatment of donor countries.[8] Furthermore, the Fund is partly capitalized through a 2% levy on the Clean Development Mechanism, which provides an independent revenue source, making the administrative apparatus more autonomous to pursue the goals prescribed. And finally, the Fund has piloted new approaches to delivering climate finance through direct access, which reduces the costs and complexity of accessing finance and helps to build capacity of receiving countries to manage adaptation.[9] Some of these progressive features have been taken up by the Green Climate Fund (GCF) and were used to reform the Global Environmental Facility (GEF).[10] Thus, the experience of the AF has informed subsequent development of adaptation finance instruments.

Governance

In order to make a brief review of the merits of the Adaptation Fund governance structure, and its ability to fulfil mandate under the Kyoto Protocol 'to assist developing country Parties that are particularly vulnerable to the adverse effects of climate change to meet the costs of adaptation',[11] it is worth setting the analysis in the context of principle-agent theory as applied to international organizations.

States may use international organizations (IO) as agents to act at arms-length where separation of global goals and local politics is needed and to help fulfil or enforce commitments.[12] Member governments (principle) having agreed on a collective action (to finance adaptation), 'hire' an organization (a dedicated team at the GEF to provide secretarial services, and World Bank trustee services) as agent to pursue the agreed goals. Agency allows governments to centralize cooperative action through concrete organizational structure, and, as such, increase efficiency, reduce transactional costs and mitigate information asymmetries. To ensure legitimacy, agents enjoy a relative independence from individual country interests and pursue their mandate within the collectively prescribed limits.[13] But principle-agent relationship does not come without challenges. The main theoretic contention is the question of control: how does the principle allow the agent to work autonomously to achieve efficiency gains without losing control?

Tools of control

Some theorists believe that interests of principle and agent never completely coincide. International agents are interested in 'the survival and growth of their organization and use their resources to further these aims'.[14] Barnett and Finnermore went as far as to say that IOs are sort of pathological 'Frankensteins' that are prone to 'dysfunctional behaviour' and undermine the interests of the states that created them.[15] However, in 2006 Nielsen et al. proposed a more balanced view, based on logic derived from agency theory in microeconomics. It is suggested that states engaging an IO as agent are far from powerless and have a number of tools to retain and exercise their control by (1) screening and selecting an agent who has similar interests, (2) monitoring the agent's performance, (3) creating a system of reward and punishment that reflects the goals of the mandate and by (4) devising a system of checks and balances that requires coordination and competition between two or more agents.[16] The institutional framework of AF reflects most of the above mentioned safeguards.

First, following the Decision 1/CMP.3, the GEF provides secretariat services to the Adaptation Fund Board through a memorandum of understanding in a 'functionally independent and effective manner'.[17] The World Bank serves as the interim trustee under terms and conditions that have been amended several times in accordance with the applicable provisions of the World Bank's Articles of Agreement, by-laws, policies and procedures.[18] Governments already have a relationship with the named organizations, including in the areas of managing environmental and climate funds, and as such, the experience, institutional memory and goals of the those IOs and the new financial instrument coincide.[19] Employing the same agent (or one derived from the other) creates a greater institutional dependency and creates more financial leverage for the principle. Nielson et al. provide an example where environmental reform of the World Bank which was at odds with lending preferences of the Bank was partly achieved by US Congress threatening to withhold replenishment of IDA and an increase of capital of the International Bank for Reconstruction and Development (IBRD).[20]

The drawback of the current set up with the GEF is the lack of control the AFB has over staff, who are remunerated and incentivized through general scheme in the GEF. As a result, there is no specific reward and punishment system that would reflect the goals of the AFB mandate. This is even more evident with the World Bank serving as a trustee, where there is a lack of accountability in managing or clarity with respect to the liability for any credit losses incurred on AF funds under IBRD management.[21]

Second, the performance of the administrative apparatus is regularly appraised by regular reviews.[22] 2011 the Review of the Interim Arrangements of the Adaptation Fund included SWOT analysis and among other things made recommendations to reinforce organizational independence of the secretariat, and hire experts as permanent staff, enhance transparency and accountability of its operations. These recommendations were duly followed by updating the operations guidelines, introducing transparency procedures.[23] Indeed, arrangements were made to make fund more independent. The latest Third Review of the Adaptation Fund looked at the effectiveness, sustainability and adequacy of the fund and its operations and concluded that the secretariat and trustee function effectively and sustainably and that the cost is mid-ranged compared to other funds.[24]

Third, there is certainly competition. In recent decades there has been a proliferation of multilateral funds providing climate finance with overlapping roles and duplication of efforts.[25] AF is only one among many funds that deal with adaptation (e.g. the Least Developed Countries Fund (LDCF), the Special Climate Change Fund (SCCF) have mandates in resilience). Moreover, the aim of Green Climate Fund is to become the main global fund for climate change and provide both adaptation and mitigation finance. As such, there is competition, if not a threat for the Adaptation Fund to be absorbed by the GCF.[26] At the same time, monetization of the certified emission reductions (CERs) failed to ensure adequate and predictable revenue streams for the AF[27] (the prospects of monetizing of ERUs and AAUs levy is too early to call) and as such, just like other multilateral funds, AF competes for donor pledges, forcing it to specialize, be more efficient and effective in order to attract donations and retain its place in the plethora of multilateral climate funds.

This brief analysis of the principal's control tools over the agent reveal that the Kyoto Protocol Parties ensured the selection of an agent with similar interests by appointing the GEF and World Bank to important functions in the operationalization of the AF. Furthermore, it is clear that monitoring of the agent's performance is not only a paper prescription but also is implemented, and at least some recommendations are taken on board and there is plenty of competition. No wonder the latest appraisal of the Fund concluded 'that the trustee and the secretariat to the Board have continued to function effectively and sustainably'.[28] What the technical review and the above analysis fail to reveal, however, is the extent of the classic fallacy that plagues principal – agent relationships in international governance contexts, and that is of capacity of the supervisory board to use tools of control at their disposal.

Agency slack

There are two characteristics of the principle-agent relationship in international governance that make it particularly prone to difficulties: the complexity of the principle, and the length of delegation chain. The classical view of the delegation chain sees the ultimate principle being the voter of the state which is a party to an international agreement, which the agent serves. As the example of the AFB chain of delegation from the perspective of a European Union citizen, see illustration 1 below that was devised on the basis of Vaubel's flow chart of the principal–agent problem of international organizations.[29] This shows that the secretariat of the Fund, which performs daily tasks of accreditation, selection of institutions, programmes, and the trustee which manages funds and performs CER monetization, are far removed from the voters of the Kyoto Protocol Parties. The staff of the agent is not rewarded for responding to demands other than those of the immediate principal (the AF Board members).[30] This creates 'agency slack'. And the more links in the delegation chain there are, the more slippage occurs and less likely it is that the interests of the ultimate principle – the citizen – are reflected. In other words, voters might not be getting from the membership what they 'paid' for.[31]

Capacity of the Adaptation Fund board

The crucial link in the delegation chain that can minimize the effects of the agency slack and ensure representation of KP Parties views is the AF Board. The Board is the operating entity of the Fund. 'It supervises and manages the Adaptation Fund under the authority and guidance of the CMP.'[32] Among functions prescribed to the Board is strategic planning, development of operations policies, monitoring and reviewing of implementation of those operations, development of legal and administrative arrangements for the secretariat. The meetings of the Board also serve as a place for principle to interact with the secretariat's staff and experts to provide 'political legitimacy to technical judgement' as well as serve as the forum where representors interact between themselves. The Board is meant to perform all four roles of supervisory body as envisaged by Martinez-Diaz, i.e. that of (1) Performance police, (2) Strategic thinker, (3) Political counterweight and (4) Democratic forum.[33] (He devised a list of indicators for measuring capacity of a board to play those roles which is useful in assessing Adaptation Fund Board functionality.)

It is argued that small boards of 12 or less are more effective in creating 'an environment of frank and constructive deliberations' and thus better suited to play a role as a strategic thinker.[34] The Adaptation Fund Board is composed of 16 members and 16 alternates representing Parties to the Kyoto Protocol. A majority of members – about 69% – represent developing countries.[35] Considering the vast membership of the Kyoto Protocol, consisting of 191 states (189 United Nations member states as well as the Cook Islands and Niue) and one supranational union (the European Union), the Board is relatively small.[36]

However, the trade-off of a small Board is that states are represented via large constituencies.

In a multi-country constituency setting, gaps in representation occur, particularly where constituency encompasses both contributors as well as receivers of adaptation funds. The Eastern European Group is an example. The region includes Annex I Parties of the Convention like Poland and non-Annex I Parties as such as Armenia. Their interests might be at odds. For example, one might have a vested interest in keeping the number of adaptation financing opportunities high, while the other, in the name of efficiency of financial mechanisms, might desire an incorporation of the Adaptation Fund into the GCF structure.

Rotation within the constituency should provide more opportunities for direct representation of individual countries. However, a survey of The AF Board Membership History Chart[37] reveals very limited rotation. In the past 10 years, Norway and Switzerland served five terms as members and as alternates despite the rule that 'members and alternate members shall serve for a term of 2 years and shall be eligible to serve a maximum of two consecutive terms' [38] by skipping a term and re-entering the Board. As a comparison, representative of Germany – by far largest contributor of the Fund – only served two terms.[39] True, long service means that Board members build institutional knowledge and expertise, arguably enabling them to better supervise the secretariat and trustee. It also strengthens the ability of a long serving Board member to perform the role of strategic thinker (for a comparative case within the overall UNFCCC framework, see the chapters by Kennedy and Krause concerning the Technology Executive Committee and the Advisory Board of the Climate Technology Centre and Network). However, this set-up undermines the legitimacy of the Board as a democratic forum.

This gives rise to the question of how far one country representing their multi-country constituency is able to reflect the interests and views of the whole region, considering that there are generally no formal consultation systems among the members of the same region (the EU being the obvious exception). This fallacy of Board membership rotation creates an environment for Board members to 'go native' and side more closely with management and staff of the agent rather than upholding the interests of the states the Board members represent. And while there may be many reasons why rotation of representation is weak, it raises the question of whether the Board can truly play a role of political counterweight to the secretariat and the trustee.

The role of the Board is further weakened by limited 'face time'. The Board only meets twice a year and thus some decisions are made inter-sessionally, limiting the ability to debate the operations of the AF.[40] And while it is expected for Board members, including alternate members, to have an 'appropriate technical, adaptation and/or policy expertise',[41] the reality is that CVs of Board members are rarely appraised due to the political nature of appointments in UN bodies.[42]

The adoption of the AF's Medium-term Strategy documents illustrate the nature and limitations of Board oversight, which are apparent from the history of strategy drafting: the Board requested the secretariat to prepare 'options for a framework for a medium-term strategy for the Fund', the secretariat subsequently recruited an external consultant to help draft it, the Board devised a Task Force to supervise and guide the drafting, the draft underwent a public consultation where NGOs and other stakeholders were invited to comment on the draft, before with corrections it was adopted at the 30th AFB meeting.

The analysis shows that there are clear trade-offs while balancing different roles of the Board. A small number of Board members makes decisions more effective, but at the expense of political counterweigh and a genuinely functioning democratic forum. Thus strengthening of the Board membership rotation rules and the calibre of Board members might be desirable in the future.

Conclusion

As adaptation gains more prominence among the global responses to climate change and UNFCCC debates concerning the future arrangements of adaptation support continue, it is useful to reflect on the nature of relationship between the Partiers of the UNFCCC and the agents they employ to implement operations. A brief review of AF institutional arrangements shows that while the Board which is tasked with supervision and management of the Fund's operations has tools to control and oversee the secretariat and the trustee as an agent, it potentially lacks capacity to fulfil its duty. And while there is no reason to believe the secretariat or the trustee undermine the interests of the Parties, the Board nonetheless fails on its key function to supervise the agent and reflect the views of principle – the voters of the Kyoto Protocol Parties.

It might therefore desirable to revise the rules of appointment and rotation of the Board members to reflect closer the technical expertise needed for Board members to perform their duties. Furthermore, the Board might benefit from permanent rather than ad-hoc experts/non-executive directors who have necessary expertise to provide independent advice on daily operations of the trustee and secretariat alongside the Board members. These points may be generalized for the improved functioning of other constituted bodies under the UNFCCC, particularly as the regime shifts focus from negotiation to implementation.

Notes

1 Art. 6.4. 'Parties recognize that the current need for adaptation is significant and that greater levels of mitigation can reduce the need for additional adaptation efforts, and that greater adaptation needs can involve greater adaptation costs.' (UNFCCC, 2015).
2 'The increases in the atmospheric concentrations of greenhouse gases enhance the natural greenhouse effect and that this will result on average in an additional warming of the Earth's surface and atmosphere' and that where 'there are threats of serious or irreversible damage' policies and measures should be comprehensive: not only cover all relevant sources, sinks and reservoirs of greenhouse gases but also adaptation. UNFCCC (United Nations 1992).
3 Art. 2. Paris Agreement (UNFCCC 2015).

4 The Agreement, among other things, calls for Art. 7 'Parties should strengthen their cooperation on enhancing action on adaptation', Art. 9.4. 'The provision of scaled-up financial resources should aim to achieve a balance between adaptation and mitigation, taking into account country-driven strategies, and the priorities and needs of developing country Parties, especially those that are particularly vulnerable to the adverse effects of climate change and have significant capacity constraints, such as the least developed countries and small island developing States, considering the need for public and grant-based resources for adaptation.' (UNFCCC 2015).
5 The Future of the Funds (World Resources Institute 2017).
6 See The Future of the Funds (World Resources Institute 2017), The effectiveness of climate finance: a review of the Adaptation Fund (ODI 2013), Power, Responsibility, and Accountability: Re-Thinking the Legitimacy of Institutions for Climate Finance (World Resource Institute 2010), Global Landscape of Climate Finance 2017 (Climate Policy Initiative 2017).
7 Decision 1/CMP.3 (UNFCCC 2008).
8 Delegation to International Organizations: Agency Theory and World Bank Environmental Reform (Hawkins 2006; Nielson, 2003).
9 The effectiveness of climate finance: a review of the Adaptation Fund (ODI 2013).
10 Power, Responsibility and Accountability: Re-Thinking the Legitimacy of Institutions for Climate Finance (World Resource Institute 2010).
11 Art. 12.8, Kyoto Protocol (UNFCCC 1998).
12 Martinez-Diaz 'Executive Boards in International Organizations: Lessons for Strengthening IMF Governance' (Martinez-Diaz 2008).
13 Why States Act through Formal International Organizations (Abbott 1998).
14 Principal–Agent Problems in International Organizations (Vaubel 2006).
15 The Politics, Power, and Pathologies of International Organizations (Barnett 1999)
16 Delegation to International Organizations: Agency Theory and World Bank Environmental Reform (Nielson and Tierny 2003).
17 Decision 1/CMP.3 (UNFCCC 2008).
18 Ibid.
19 While the Review of the Interim Arrangements of the Adaptation Fund (Adaptation Fund, 2011) noted that considering the maturity of the Fund, it could be beneficial to separate GEF and AF secretariat, it nonetheless points out to Tanzania example, where operational linkages between the GEF and the Adaptation Fund helped reduce costs, build synergies and avoid duplication while implementing AF and LDCF projects in the same area of intervention. Furthermore, the Third review of the Adaptation Fund reiterated the complementarity of institutional arrangements between the Fund board and secretariat and interim trustee: 'the Adaptation Fund benefits from its secretariat being housed in the GEF secretariat, in terms of the complementarity between the different funds and between the Adaptation Fund Board and GEF secretariats' (Adaptation Fund 2017).
20 See Delegation to International Organizations: Agency Theory and World Bank Environmental Reform (Nielson and Tierny 2003).
21 Review of the Interim Arrangements of the Adaptation Fund (Adaptation Fund 2011).
22 Decision 1/CMP.3 (UNFCCC 2008).
23 Operational Policies and Guidelines (Adaptation Fund 2017).
24 Third Review of the Adaptation Fund (Adaptation Fund 2017).
25 The Future of the Funds (World Resources Institute 2017).
26 Potential Linkages between Adaptation Fund and Green Climate Fund (Adaptation Fund 2015).
27 Third Review of the Adaptation Fund (Adaptation Fund 2017).
28 Third Review of the Adaptation Fund (Adaptation Fund 2017).

29 Principal–Agent Problems in International Organizations (Vaubel 2006).
30 As shown above, the lack of the system of reward and punishment that would reflects the goals of the mandate is a weak link in AFB principle-agent control mechanism.
31 One can argue that disconnect between AF agent and the ultimate principle is particularly apparent in the case of EU voters (and that of the countries that participate if carbon trade as well as pledge to AFB) Some country citizens contribute to the AF twice: (a) by participating in CDM, (or JI or AAU trade), the levy of which goes into the Fund, and (b) by the virtue of voluntary pledges consisting of public funds (taxpayers' money) that make up almost half of AF funds. The largest donor by far is Germany (World Bank 2017). If, as in the case of IBRD or IMF supervisory boards, where large donors yield greater power over the agent (international organization), EU citizen should have double the power over the strategy and direction of AFB, yet it is not the case. So from the perspective of the tax paying voters of the EU they are mostly removed from the actual business of financing adaptation through AF.
32 Decision 1/CMP.3 (UNFCCC 2008).
33 Executive Boards in International Organizations: Lessons for Strengthening IMF Governance (Martinez-Diaz 2008).
34 Ibid.
35 (a) Two representatives from each of the five United Nations regional groups; (b) One representative of the small island developing States; (c) One representative of the least developed country Parties; (d) Two other representatives from the Parties included in Annex I to the Convention (Annex I Parties); (e) Two other representatives from the Parties not included in Annex I to the Convention (non-Annex I Parties). An alternate is elected for each representative. Decision 1/CMP.3.
36 For comparison of board members in the international organisations see Executive Boards in International Organizations: Lessons for Strengthening IMF Governance (Martinez-Diaz 2008).
37 Adaptation Fund Board Membership History Chart (UNFCCC 2017).
38 Decision 1/CMP.3 para9.
39 This is s stark difference between composition and voting powers of AF board and other IOs. EU taxpayers yield disproportionally little control for what they pay. EU citizens contribute to the AF twice: (a) by participating in CDM, the levy of which goes into the fund, and (b) by the virtue of voluntary pledges consisting of public funds (taxpayers' money) that make up almost half of AF funds. The largest donors are Germany, Sweden and Spain. If, as in the case of World Bank or IMF boards, where large donors yield greater power over the agent (international organisation), EU citizen should have double the power over Adaptation Secretariat and the World Bank as a trustee, yet it is not the case. So from the perspective of the tax paying voters of the EU they are mostly removed from the actual business of financing adaptation through AF.
40 For list of AFB meetings and meetings' reports, see www.adaptation-fund.org/documents-publications/meeting-documents/.
41 Decision 1/CMP.3.
42 Information on membership of UNFCCC and Kyoto Protocol bodies see http://unfccc.int/bodies/election_and_membership/items/7121.php Note that use of experts is envisaged in the operations of AFB.

Bibliography

Abbott, K.W. (1998). *Why States Act through Formal International Organizations*. Retrieved from http://journals.sagepub.com/doi/abs/10.1177/0022002798042001001.

Adaptation Fund. (October 2017). *Operational Policies and Guidelines for Parties to Access Resources from the Adaptation Fund*. Retrieved from www.adaptation-fund.org/wp-content/uploads/2017/10/AFB.B.30.7_OPG-amended-in-October-2017.pdf.

Adaptation Fund. (November 5 2011). [T. Rouchdy, ed.]. *Review of the Interim Arrangements of the Adaptation Fund*. Retrieved from www.adaptation-fund.org/wp-content/uploads/2015/01/Review%20of%20Interim%20Arrangements-Final.pdf.

Adaptation Fund. (March 2015). *Potential Lineages between Adaptation Fund and Green Climate Fund*. Retrieved from www.adaptation-fund.org/wp-content/uploads/2016/05/AFB.B.25.Inf_.6-Potential-linkages-between-AF-GCF.pdf.

Adaptation Fund. (2017). *Third Review of the Adaptation Fund*. Retrieved from http://unfccc.int/resource/docs/2017/tp/06.pdf.

Barnett, M.N. (1999). "The Politics, Power, and Pathologies of International Organizations." *International Organization*, 699–732.

Climate Policy Initiative. (2017). [P.O. Barbara and K. Buchner, eds]. *Global Landscape of Climate Finance 2017*. Retrieved from https://climatepolicyinitiative.org/wp-content/uploads/2017/10/2017-Global-Landscape-of-Climate-Finance.pdf.

Hawkins, D.G. (2006). *Delegation and Agency in International Organizations*. Cambridge University Press, Cambridge, UK.

IPCC. (November 1 2007). *Climate Change 2007: Impacts, Adaptation and Vulnerability. Contribution of Working Group II to the Fourth Assessment Report of the Intergovernmental Panel on Climate Change*. Retrieved from www.ipcc.ch/pdf/assessment-report/ar4/wg2/ar4_wg2_full_report.pdf.

IPCC. (2014). *Climate Change 2014: Synthesis Report. Contribution of Working Groups I, II and III to the Fifth Assessment Report of the Intergovernmental Panel on Climate Change*. IPCC, Geneva, Switzerland. Retrieved from www.ipcc.ch/pdf/assessment-report/ar5/syr/SYR_AR5_FINAL_full_wcover.pdf.

Martinez-Diaz, L. (2008). Executive Boards in International Organizations: Lessons for Strengthening IMF Governance. *Global Economy and Development Program*. Retrieved from www.ieo-imf.org/ieo/files/completedevaluations/05212008BP08_08.pdf.

Nielson, D.L. and Tierny, M.J. (2003). "Delegation to International Organizations: Agency Theory and World Bank Environmental Reform." *International Organization*, 241–276.

ODI. (2013). [T. N. S., ed.]. *The Effectiveness of Climate Finance: A Review of the Adaptation Fund*. Retrieved from www.odi.org/publications/7381-multilateral-climate-finance-effectivenessadaptation-fund-unfccc.

UNFCCC. (1998). *The Kyoto Protocol to the UNFCCC*. Retrieved from http://unfccc.int/resource/docs/convkp/kpeng.pdf.

UNFCCC. (March 2008). *Decisions Adopted by the Conference of the Parties serving as the Meeting of the Parties to the Kyoto Protocol*. Retrieved from http://unfccc.int/resource/docs/2007/cmp3/eng/09a01.pdf#page=3.

UNFCCC. (2015). *Paris Agreement*. Retrieved from http://unfccc.int/files/meetings/paris_nov_2015/application/pdf/paris_agreement_english_.pdf.

UNFCCC. (2017). *Adaptation Fund Board Membership History Chart*. Retrieved from https://unfccc.int/files/bodies/election_and_membership/application/pdf/consolidated_historical_membership_charts_of_convention,_kp_and_cma_bodies.pdf.

United Nations. (1992). *Framework Convention on Climate Change*. Retrieved from https://unfccc.int/resource/docs/convkp/conveng.pdf.

Vaubel, R. (2006). "Principal–Agent Problems in International Organizations." *International Organizations*, 1, 2, 125–138.

World Bank. (September 30 2017). Retrieved from Adaptation Fund Trust Fund Financial Report: http://fiftrustee.worldbank.org/SiteCollectionDocuments/ADAPT/Trustee%20 Reports/AF_TR_09_17.pdf.

World Resources Institute. (2010). [S.N. Athena Ballesteros, ed.]. *Power, Responsibility and Accountability: Re-Thinking the Legitimacy of Institutions for Climate Finance*. Retrieved from www.wri.org/publication/power-responsibility-and-accountability.

World Resources Institute. (2017). [J.T. Niranjali Amerasinghe, ed.]. *The Future of The Funds*. Retrieved from www.wri.org/sites/default/files/The_Future_of_the_Funds_0.pdf.

12 The climate–energy–trade nexus in EU external relations

Rafael Leal-Arcas[1] *and Eduardo Alvarez Armas*

Introduction

There is a lot of talk about the fact that, as a result of trade, we have increased social inequality nationally and internationally and that the level of carbon dioxide and other greenhouse gas (GHG) emissions has been going up over time, also a result of international trade,[2] and what can be done about it.[3] Trade is in many senses considered to be the nemesis of environmental protection. In this chapter, however, we argue that, while all of that may be true, trade can contribute to climate change mitigation.

There is no question that climate change is one of the biggest challenges humanity faces today. Today, 80% of the global energy supply comes from fossil fuels.[4] Fossil fuels contribute to climate change and are finite, which leads to energy insecurity. Renewable energy can help here in that it is cleaner than fossil fuels. It also helps towards energy independence and therefore enhances energy security. Trade law could be used as a vehicle to achieve this goal. We argue that trade agreements can help mitigate climate change since, in the past, they have been a very powerful instrument for change, as the following two examples demonstrate:

1. poverty reduction: due to trade agreements, one billion people have come out of poverty between 1990 and 2010;[5] and
2. the protection of human rights: 75% of countries use trade agreements to protect human rights.[6]

So why not use trade agreements as a novel tool to solve one of the most important challenges of today, namely climate change?

The purpose of this chapter is not to thoroughly describe the current state of the climate–energy–trade nexus in the external relations of the European Union (EU), but to explore, in normative terms, the potential of the EU's external trade[7] towards contributing to global decarbonization. As the EU has done in the case of human rights, EU free trade agreements (FTAs) may be used to help decarbonize the economy by bundling the fate of these agreements and international climate-change obligations, i.e., the breach of the latter would give rise to legal consequences under the dispute settlement mechanisms of the former (in other words, there would be no free-trade-benefits without effective fulfilment of international climate-change-related obligations). The feasibility of this idea will be assessed via the textual analysis of the current wording and content of environmental/climate-change-related elements in a sample of relatively recent and/or upcoming FTAs.

This chapter first sets the global scene on the links between energy and climate change, trade and energy as well as climate change and trade. It then provides a textual analysis[8] of environmental or climate-change-related provisions in several EU FTAs. Finally, it offers a case for bundling environmental and trade agreements before it concludes.

Setting the global scene: a triangulation to sustainable development

Linking energy and climate change[9]

Starting from the premise that climate change mitigation is a global public good, there is a nexus between energy and climate change, which encompasses a range of issues such as clean energy subsidies,[10] carbon taxes,[11] and border adjustment for carbon emissions.[12] This last point of border carbon adjustment may be a way forward in tackling climate change. As climate change is one of the most important public policy issues facing countries around the world, countries are adopting various policies in order to address these concerns. Of these, limiting anthropogenic (man-made) GHG emissions is a significant mitigation measure.

In the view of Dieter Helm, 'since 1900, the global population has more than tripled and the consumption of energy (largely fossil fuels) has increased more than tenfold. Climate change has been caused by the way resources have been consumed, and climate change policy necessitates a substantial change in the allocation of resources.'[13] Moreover, according to Kleymeyer, 'the energy sector, including energy use and production, accounts for over 50% of global [greenhouse gases] GHGs.'[14] Given rapidly rising industrialization in the developing world, and the fact that low-cost energy options are likely to be heavily fossil fuels based for some time to come, GHG emissions are projected to increase and climate change mitigation will remain an urgent issue.[15] Furthermore, when dealing with biofuels, it is necessary to find a balance between climate change and energy security concerns.[16]

Since the use of fossil fuel is one of the major sources of anthropogenic GHG emissions, it is important to promote climate-sensitive energy policy that would help countries increase non-fossil fuel sources in their energy mix.[17] Various alternative energies, in which nuclear and renewable sources play key roles, are being explored and developed by countries as part of their diversification efforts.[18] Major investments in the new and renewable energy sector will be required in order to increase non-fossil energy usage.[19]

The increased competition for energy resources, climate change and GHG emissions controls, technological advances and limitations have all contributed to a contradictory, fragmented regulatory web. These include the exploration of new sources of energy, the transition to greener resources and intelligent grids, the challenges to the security of supply networks, affordability and its links with development and the contested consumption paradigms, the nature and size of energy companies, and the cross-jurisdictional terrain on which they compete.

Linking trade and energy[20]

The presumption is that trade liberalization will increase economic activity[21] and therefore energy consumption.[22] All countries require energy resources, but few possess these, and thus trade in energy (primarily oil) is crucial to fulfil global energy needs.[23] Internationally, there is more trade in oil than in anything else. 'Fully half of world trade in services is intensely energy-dependent.'[24] Yet, the General Agreement on Tariffs and Trade/World Trade Organization (GATT/WTO) has historically not preoccupied itself with energy trade. Very few energy-rich countries saw a need to join the GATT/WTO club, given that the reduction of import restrictions – one of the main goals of the multilateral trading system – is not an issue when it comes to energy. Saudi Arabia, the main energy-producing country in the world, only joined the WTO in 2005.[25] Russia joined in 2012[26] and several energy-producing countries[27] are still not WTO Members.[28]

All forms of energy should be subject to the same rules. Energy may become part of the WTO agenda in the near future.[29] Given that current WTO rules are far from addressing all the needs of energy trade today, is it necessary to have a WTO Agreement on trade in energy?[30] If so, can and should the Energy Charter Treaty be used as a model? Moreover, since Russia finally joined the WTO on its own in 2012[31] (and not as a customs union along with Belarus and Kazakhstan)[32] and since energy is one of its greatest assets in economic terms, would this be the right time to include energy trade as part of the WTO Agreements? Those energy-rich Middle Eastern countries that are not yet WTO Members, but wish to become WTO Members, will most likely follow Russia.[33] These Middle Eastern countries should prioritize the conclusion of negotiations to enter the WTO in order to fully integrate into the global trading system and protect their growing interests on world markets. WTO membership will certainly help eliminate any discrimination against them in their trade and investment.

Linking climate change and trade[34]

Given that much of the world's energy needs are likely to largely depend on fossil fuel based sources for the near to mid future, an increase in economic activity and energy consumption in the future, together with an increase in world population, will lead to higher levels of GHG emissions. This means that there is an inevitable link between international trade and future global climate change agreements,[35] which could lead to a potential conflict between the objectives of the Kyoto Protocol/Paris Agreement on Climate Change and the WTO rules.[36] Moreover, climate change is one of the most relevant international issues facing the world today. The relationship between trade agreements and measures to mitigate the effects of climate change remains a polemic part of the debate.[37] It is therefore relevant to explore key legal and economic issues arising from the climate change debate, including the relationship between the WTO Agreement and multilateral environmental agreements that address climate change.[38]

Based on the premise that regional trade agreements (RTAs) can be used as building blocks for multilateralism, however, one could envisage a global climate agreement based on climate-related RTAs, especially large RTAs such as the Trans-Pacific Partnership.[39] Indeed, given how pro-active developing countries are in the conclusion of RTAs, this option would be an effective way toward a future global climate agreement. In this sense, RTAs can be used as a legal mechanism to move forward the multilateral environmental agenda. This avenue could be very promising given the exponential increment in the conclusion of preferential trade agreements (PTAs)/FTAs/RTAs in recent years and the increasing number of environmental provisions per FTA as depicted in Figure 12.1.[40] In fact, around 45% of climate contributions submitted to the Conference of the Parties 21 in 2015 make reference to trade measures.[41] This fact shows the increasing interaction between climate and trade.

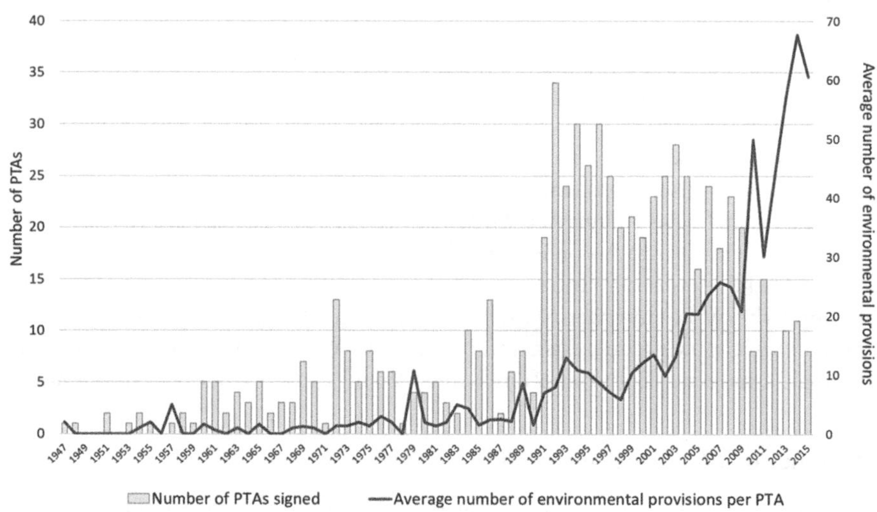

Figure 12.1 History of PTAs with environmental provisions

Source: J-F Morin et al., "Environmental Clauses in Trade Agreements: Empirical evidence for sound policies."

'Environmental' language and content in EU Free Trade Agreements

The EU leads in the number of climate-related provisions in FTAs concluded throughout the world.

When analyzing the environmental and/or climate-change-relevant substantive content of several EU FTAs and its potential articulation with dispute settlement mechanisms, different configurations can be identified therein. Let us assess them in sequence.

First configuration – ambiguous substantive content and non-enforceability: various EU Free Trade Agreements

The first possible configuration that may be found within the sample of agreements is embodied in the following five: the 'Free trade Agreement between the European Union and its Member States, of the one part, and the Republic of Korea, of the other part' (EU–South Korea FTA);[42] the 'Trade Agreement between the European Union and its Member States, of the one part, and Colombia and Peru, of the other part' (EU–Colombia–Peru TA);[43] the 'Association Agreement between the European Union and its Member States, of the one part, and Ukraine, of the other part';[44] the 'Comprehensive Economic and Trade Agreement (CETA) between Canada, of the one part, and the European Union and its Member States, of the other part' (CETA);[45] and the 'Japan-EU Economic Partnership Agreement' (JEEPA).[46] Overall, the environmental language and content of these EU FTAs seem to be relatively stable (with the notable exception, as it will be seen, of the EU–Colombia–Peru TA), and, therefore, they will be assessed simultaneously. In terms of substantive content, generally speaking, these agreements tend to contain a chapter on 'Trade and sustainable development' or 'Trade and environment', within which an article on 'Multilateral environmental agreements' may be found.[47] The content of the said article has a constant basis, upon which other elements are pinned. The said basis is constituted by the following language:

> 1. The Parties *recognize the value* of international environmental governance and agreements as a response of the international community to global or regional environmental problems.
> 2. The Parties *reaffirm their commitments* to the effective *implementation* in their laws and practices of the multilateral environmental agreements to which they are party.[48]
>
> [Emphasis added]

This common basis is sometimes accompanied by further paragraphs which, despite depicting progressive intentions, are unable to lead to ambitious environmental outcomes. As brief examples, in the agreements with South Korea and Japan, the parties mention the United Nations Framework Convention on

Climate Change and the Kyoto Protocol (or, in the case of the JEEPA, the Paris Agreement), but only to state that they '*reaffirm their commitment to reaching* [the Agreements'] *ultimate objective*'. The Agreement with Ukraine establishes further means through which the parties '*shall cooperate*', as does CETA, which adds that the parties '*commit to consult*' and '*exchang*[e] *information*'. Equally, the first paragraph above is sometimes accompanied by additional elements which remain within the same vague narrative (the parties '*commit to consulting and cooperating*', '*stress the need to enhance*', etc.).

However, beyond these considerations, which would tend to sustain the conclusion that the substantive content of the concerned Agreements is purely and simply weak, a qualification may be made: the second paragraph is ambiguous in its content. The paragraph begins with a typically 'weak' wording ('*The Parties reaffirm their commitments*'), but, instead of referring to weak policy objectives, it refers to the '*effective implementation in their laws and practices . . . of the multilateral environmental agreements to which they are party*.'[49] It, therefore, could be argued that, despite its beginning, the paragraph actually links, in a solid manner, the trade agreement to the relevant multilateral environmental agreements (any relevant agreement to which they are party, without limitation). In other words, it could be argued that a lack of '*effective implementation in their laws and practices*' of the relevant environmental agreements entails a breach of paragraph 2.

This potential interpretation is further reinforced by the specific wording of the relevant provisions of the EU–Colombia–Peru TA. The second, third and fourth paragraphs of Article 270 ('*Multilateral Environmental Standards and Agreements*'), included in Title IX ('*Trade and sustainable development*') of the said FTA, read as follows:[50]

> 2. The Parties *reaffirm their commitment to effectively implement* in their laws and practices the following multilateral environmental agreements: the Montreal Protocol on Substances that Deplete the Ozone Layer . . ., the Basel Convention on the Control of Transboundary Movements of Hazardous Wastes and their Disposal . . . the Kyoto Protocol to the United Nations Framework Convention on Climate Change. . . .
> 3. The Trade Committee *may recommend the extension of the application* of paragraph 2 to other multilateral environmental agreements. . . .
> 4. Nothing in this Agreement shall limit the right of a Party to *adopt or maintain measures to implement* the agreements referred to in paragraph 2.[51]
>
> [Emphasis added, original footnotes omitted]

Paragraph 2 of the article contains a list with a limited number of precisely identified agreements. Notwithstanding the fact that, at its present state, this provision does not seem very prone to providing coverage to climate-change instruments,[52] it may be argued that this provision does not contain vague policy objectives and statements, but clear, precise and solid obligations. If this is the case, the above-referred parallel provisions in other EU FTAs may be interpreted accordingly, as being open-ended clauses (hence, not restrained to a limited list of agreements), containing the same kind of clear, precise and solid obligations.

However, in order for this ambiguous possibility to have any practical impact, the said provisions would need to be enforceable under the relevant '*Dispute Settlement*' provisions in the respective EU FTAs, which does not seem to be the case. While, in most cases, the dispute settlement chapters of the selected FTAs are assorted with provisions which state that their scope is general and covers the entirety of the said agreements, these provisions operate only '*unless otherwise provided*'.[53] The environmentally relevant chapters, therefore, exclude resort to dispute settlement under various formulations.[54] In general, they only provide for soft and conciliatory mechanisms, such as '*Government consultations*', or the establishment of '*panels*' or '*groups*' of experts, who draft '*initial*' and, '*final*' reports, on how to solve issues.[55] The parties then '*shall take into account the final report*', or '*inform the* [relevant entity] *of its intentions as regards the recommendations*', or '*discuss actions or measures to resolve the matter in question, taking into account the . . . report and suggestions*.'[56] The parties' courses of action are often '*monitor*[ed]', but no consequence is provided for.

Second configuration – solid substantive content, but non-enforceability: the EU–Singapore Free Trade Agreement

A second possible configuration is reflected in the text of the '*Free Trade Agreement between the European Union and the Republic of Singapore*' (EU–Singapore FTA.)[57] Chapter thirteen ('*Trade and sustainable development*') contains in Section C ('*Trade and Sustainable Development – Environmental Aspects*') an Article on '*Multilateral Environmental Standards and Agreements*' (Article 13.6) which bears a structure that greatly resembles the one described in the analysis of the previous configuration (i.e., '*ambiguous substantive content and non-enforceability*'). However, on this occasion, there is no possible ambiguity:

> 2. The Parties shall *effectively implement* in their respective laws, regulations or other measures and practices in their territories, *the multilateral environmental agreements to which they are party*.[58]
>
> [Emphasis added; original footnote omitted]

The wording of the second paragraph of Article 13.6 of the EU–Singapore FTA is clearly assertive, without any sort of resort to the typical vague language found in weak policy statements. It can hardly be argued that it does not contain clear, precise and solid obligations. *A priori*, its only fault, for the purposes of this chapter, is that it clearly does not cover climate-change obligations, for climate change is the object of the third paragraph of the Article, which is undoubtedly weak in its content:[59]

> 3. The Parties *reaffirm their commitment* to reaching the ultimate objective of the UN Framework Convention on Climate Change (hereinafter referred to as 'UNFCCC'), and of its Kyoto Protocol in a manner consistent with the principles and provisions of the UNFCCC. They *commit to work* together to strengthen the multilateral, rules-based regime under the UNFCCC building on the UNFCCC's agreed decisions, and to support efforts to develop a post-2020 international climate change agreement under the UNFCCC applicable to all parties.
>
> [Emphasis added]

In any case, again, as was the case in the first model, for the second paragraph of this provision to have any practical impact, it would need to be enforceable under the relevant '*Dispute Settlement*' chapter. And, again, this does not seem to be the case.[60]

Third configuration – enforceability, but ambiguous substantive content: the Transatlantic Trade Investment and Partnership

The '*Transatlantic Trade Investment and Partnership*' (TTIP), which the European Union and the United States began negotiating between the 7th and the 12th of July 2013, is at the roots of the third configuration detected in the sample of agreements. Notwithstanding the fact that a significant amount of uncertainty swirls around its future,[61] draft-documents on this envisaged agreement, reflecting different stages in the negotiations, will serve as the basis for the analysis. The first of such documents is the EU's '*initial proposal for legal text on "Trade and Sustainable Development"*',[62] drafted to serve during the negotiating round with the United States that took place between the 19th and the 23rd October 2015. Within '*Section III*' ('*Trade and Sustainable Development – Environmental aspects*'), Article 10 ('*Multilateral environmental governance and rules*') contains the following paragraph:

> 2. Each Party *reaffirms its commitment* to effectively *implement* in its domestic laws and practices the Multilateral Environmental Agreements to which it is a party.
>
> [Emphasis added]

Other than this paragraph, verbs and terminology in the remainder of the provision are, again, weak in nature (i.e. '*recognize*', '*continue to strive*', '*exchanging information*', '*supporting each other . . .*', '*commit to consult and cooperate*', '*acknowledge*'). Hence, in general terms, the wording of Article 10 roughly corresponds with that of equivalent substantive provisions in the EU FTAs that were classified as having an '*ambiguous substantive content*' but lacking enforceability (i.e. the first configuration above). What distinguishes this early TTIP document from the above-commented FTAs is precisely the fact that, following an EU's '*initial proposal for legal text on "Dispute Settlement (Government to*

Government)"', the TTIP's relevant provisions would be enforceable, and that remedies would be available. Article 2 ('*Scope of application*') in Section 1 ('*Objective and Scope*') establishes that the dispute settlement chapter applies to '*any dispute concerning the interpretation and application of the provisions of this Agreement, except as otherwise provided*'.

This assessment, made on the basis of a rather early-stage document, may be altered if the content of a second document, stemming from a later round of the negotiations (11–15 July 2016), gets to be inserted into the relevant chapter of the TTIP. The EU's '*initial proposal for a legal text on climate which would be included* [sic] *to the "Trade and Sustainable Development" chapter in TTIP*'[63] contains lengthy language on '*Trade favouring low-emission and climate-resilient development*' and '*Protection of the Ozone Layer and Measures Related to Hydrofluorocarbons*'. For the sake of brevity (especially taking into account the above-referred uncertainty surrounding the negotiations), suffice it to say that most of the content of the draft provisions can be easily qualified as 'weak', for it reflects the same kind of language as other provisions described in previous pages. However, each of the two unnumbered draft provisions features a paragraph bearing more solid content:[64]

> 4. . . . the Parties shall: . . . b) *effectively implement* the WTO Environmental Goods Agreement (EGA) and in this context cooperate to reduce or, as appropriate, eliminate non-tariff barriers related to environmental goods and services;[65]
>
> 2. . . . each Party *shall take measures* to control the production and consumption of, and trade in, substances within the scope of the Montreal Protocol on Substances that Deplete the Ozone Layer, including any future amendments thereto.[66]
>
> [Emphasis added]

Nevertheless, as none of these paragraphs with 'solid' content refers to any of the core climate-change agreements, they may only be collaterally relevant, if anything, for the purposes of this chapter.

The potential insertion of this language in the TTIP would likely remove climate change from the scope of the above-referred Article 10. This would mutate this third configuration, which would thus become '*enforceability, but weak content*'. It would, nevertheless, remain a distinct one, despite containing solid obligations as regards collaterally-relevant agreements: on the one hand, the provision is weak and, on the other hand, enforcement is available.

However, irrespective of the availability of enforcement, a major hindrance in this respect would always remain: putting aside the actual architecture and specific features that a final TTIP may possibly come to display, what is really fundamental is the fact that the United States has not ratified the Kyoto Protocol and may withdraw from the Paris Agreement on Climate Change.[67] If that were the case, consequently, there would be no relevant climate-change obligation that could potentially come to be enforced via this trade agreement, other than the rather weak obligations assumed under the Climate Convention itself.

The case for bundling trade and environmental agreements

Empirical research shows that, as the law stands nowadays, there are, at least, three sorts of obstacles to the hypothesis presented in this chapter:

1 Ambiguous language of the relevant 'primary' obligation: whenever a given climate-change relevant provision is covered by the dispute settlement mechanism, the actual content of the relevant obligation is ambiguous.
2 Lack of coverage by the relevant dispute-settlement provisions. This is the reverse problem: while the content/language of the relevant obligations is solid, the provision is not covered by the dispute settlement mechanism.
3 Lack of environmental commitment by at least one of the parties: no bundled climate-change enforcement would be possible because the relevant party has not ratified any meaningful international climate change obligation.

Despite this, in normative terms (*de lege ferenda*), an effective way for the European Union to move forward regarding the enforcement of international climate change mitigation obligations would be to bundle the fate of climate-change-relevant obligations with that of trade agreements. Since the relevant elements for 'bundling' are already disseminated in various FTAs (some feature solid climate change-linkage language, others feature coverage by the relevant dispute-settlement mechanism), law-makers would need to simultaneously include both sorts of elements in future FTAs to obtain the desired result:

In substantive terms, possibly the clearest wording found in the analyzed agreements is the one featured in Article 13.6 of the EU–Singapore FTA:

> 2. The Parties shall *effectively implement* in their respective laws, regulations or other measures and practices in their territories, *the multilateral environmental agreements to which they are party*.
>
> [Emphasis added; original footnote omitted]

It is not a closed or limited list of agreements, but a general clause, and leaves no doubt regarding the kind and extent of obligation being assumed.

In enforceability terms, it will suffice to follow the example of the TTIP. In other words, the relevant provision needs to be covered by the Dispute Settlement Chapter of the relevant agreement, which needs to ultimately provide access to remedies such as compensation and suspension of benefits in case of 'non-implementation' of panel reports.

All in all, a 'partially progressive' trend towards using trade agreements as legal instruments to protect the environment is possibly starting: while CETA and JEEPA face, cumulatively, ambiguous content and un-enforceability, the EU–Singapore FTA only suffers from un-enforceability, but has clearly solid content.

Conclusions

Very little is still known on the interaction between trade agreements and climate action. Yet trade is increasingly becoming an important element in the intersection with climate action. Similarly, there is an increasing interest in inserting climate-related provisions in the WTO context and in FTAs. Reference to climate change, the UNFCCC, the Kyoto Protocol and the Paris Agreement in FTAs in increasing. However, the empirical analysis in this chapter has shown that there are still limitations to how trade agreements can contribute to climate change mitigation. This denotes that there is still a major gap between the trade and climate communities to answer questions such as how can one reduce GHG emissions from energy-intensive traded goods without economic loss?

Being creative in legislative terms and accepting the fact that most international trade rules were written for a 20th century reality should be enough reasons to be optimistic that we may soon see trade agreements that address 21st century challenges such as climate change and clean energy technologies. In fact, evidence of this trend is the fact that recent bilateral and regional trade agreements cover, albeit in a light manner, climate-related issues by inserting chapters in such agreements that refer to the links between trade and climate change. Therefore, as time progresses and the science of climate change becomes more robust, there is a great chance that future free trade agreements may incorporate strong chapters on climate-related issues. The EU is perfectly placed to exploit this nexus within the realm of its external relations.

Finally, a 'peace clause' for climate action in future (EU) FTAs could be a way forward in finding supportive avenues between trade and climate action. Similar to the WTO's Agriculture Agreement, the peace clause would enable countries to protect themselves from action taken against each others' fossil fuel subsidies. The trading system could also be more accommodating to climate objectives if countries reduced tariffs for environmental goods and services and removed or reduced fossil fuel subsidies. Such actions would contribute greatly to GHG emissions reduction. In any event, environmental protection should not be lowered in the name of trade liberalization or investment promotion.

Notes

1 The financial help from two EU grants is greatly acknowledged: Jean Monnet Chair in EU International Economic Law (project number 575061-EPP-1–2016–1-UK-EPPJMO-CHAIR) and the WiseGRID project (number 731205), funded by Horizon 2020. Both grants have been awarded to Professor Dr. Rafael Leal-Arcas.

2 *Is Trade Good or Bad for the Environment?* OECD, www.oecd.org/trade/tradeand environment.htm (last visited 17 July 2017) [https://perma.cc/4VXK-KTWC]; see also *Environmental Goods and Services Sector*, Europa: Eurostat, http://ec.europa.eu/eurostat/web/environment/environmental-goods-and-services-sector (last visited 17 July 2017) ("The purpose of environmental goods and services is to prevent, reduce and eliminate pollution and any other form of environmental degradation . . . and to conserve and maintain the stock of natural resources, hence safeguarding against depletion.") [https://perma.cc/5R27-B7FH].

3 Some policy suggestions include a tax on the carbon content of imports and refund the tax to companies when they export, as the European Union is doing with cement. See The Economist, "Externalities: The Lives of Others," 19 August 2017, pp. 60–61, at 61.

4 *World Energy Council Report Confirms Global Abundance of Energy Resources and Exposes Myth of Peak Oil*, World Energy Council (15 October 2015), www.worldenergy.org/news-and-media/press-releases/world-energy-council-report-confirms-global-abundance-of-energy-resources-and-exposes-myth-of-peak-oil/ [https://perma.cc/5ZK6-8LNL].

5 See *Towards the End of Poverty*, The Economist (1 June 2013), www.economist.com/news/leaders/21578665-nearly-1-billion-people-have-been-taken-out-extreme-poverty-20-years-world-should-aim [https://perma.cc/WHD7-8TZY].

6 Susan Ariel Aaronson, *Human Rights*, The World Bank 443, 443 http://site resources.worldbank.org/INTRANETTRADE/Resources/C21.pdf (last visited 17 July 2017) [https://perma.cc/P5KD-KZX8].

7 For a comprehensive analysis of EU external trade law and policy, see Leal-Arcas, R. *Theory and Practice of EC External Trade Law and Policy*, Cameron May, 2008; Leal-Arcas, R. *International Trade and Investment Law: Multilateral, Regional and Bilateral Governance*, Edward Elgar, 2010.

8 Due to word-limit constraints, for the purposes of this chapter, we focus solely on a textual analysis of free trade agreements.

9 See generally Leal-Arcas, R. *The European Energy Union: The Quest for Secure, Affordable and Sustainable Energy*, Claeys & Casteels Publishing, 2016.

10 Cosbey, A. "Revenue Energy Subsidies and the WTO: The Wrong Law and the Wrong Venue," Global Subsidies Initiative, available at www.globalsubsidies.org/subsidy-watch/commentary/renewable-energy-subsidies-and-wto-wrong-law-and-wrong-venue; see also Bigdeli, S., "Incentive Schemes to Promote Renewables and the WTO Law of Subsidies," in Cottier, T., Nartova, O., and Bigdeli, S. (eds), *International Trade Regulation and the Mitigation of Climate Change*, New York, Cambridge University Press, 2009, pp. 155–192; Howse, R. "Climate Mitigation Subsidies and the WTO Legal Framework: A Policy Analysis," Geneva, Switzerland, International Institute for Sustainable Development, 2010.

11 Sindico, F. "Climate Taxes and the WTO: Is the Multilateral Trade Regime a Further Obstacle for Efficient Domestic Climate Policies?" *EcoLomic Policy and Law*[MS1], Vol. 3, No. 7/8, 2006, pp. 116–139.

12 Kaufmann, C. and Weber, R. "Carbon-related Border Tax Adjustment: Mitigating Climate Change or Restricting International Trade?" *World Trade Review*, published online on 16 August 2011; Burniaux, J.-M., Chateau, J. and Duval, R. "Is There a Case for Carbon-Based Border Tax Adjustment?: An Applied General Equilibrium Analysis," *Economics Department Working Papers*, No. 794, 2010; Holmes, P., Reilly, T. & Rollo, J. "Border Carbon Adjustments and the Potential for Protectionism," *Climate Policy*, Vol. 11, No. 2, 2011, pp. 883–900; Genasci, M. "Border Tax Adjustments and Emissions Trading: The Implications of International Trade Law for Policy Design," *Carbon and Climate Law Review*, Vol. 2, No. 1, 2008, pp. 33–42; Ismer, R., and Neuhoff, K., "Border Tax Adjustment: A Feasible Way to Support Stringent Emission Trading," *European Journal of Law and Economics*, Vol. 24, 2007, pp. 137–164; de Cendra, J. "Can Emissions Trading Schemes be Coupled with Border Tax Adjustments? An Analysis vis-à-vis WTO Law," 15 *Review of European Community & International Environmental Law*, 2006.

13 Helm, D., (ed.) *Climate-Change Policy* (Oxford University Press, 2005), pp. 12–13.

14 Kleymeyer, A.M., "New Rules for the Environmental Imperative: Considerations for the Energy Sector and Interaction with WTO Rules," in Pauwelyn, J. (ed.), *Global Challenges at the Intersection of Trade, Energy and the Environment*, Geneva, Switzerland, Centre for Trade and Economic Integration, 2010, p. 63.

15 For an estimate of CO2 emissions by country from 1971 to 2008, see International Energy Agency Statistics, *CO2 Emissions from Fuel Combustion: Highlights*, Paris: OECD/IEA, 2010.

16 On biofuels, see the views of Searchinger, T. et al. "Use of U.S. Croplands for Biofuels Increases Greenhouse Gases Through Emissions from Land-Use Change," *Science*, Vol. 319, No. 5867, 2008, pp. 1238–1240; Fischer, C., Eggert, H. and Mavroidis, P. "The Law and Economics of Trade in Biofuels," ENTWINED Issue Brief, 2010/16/09.

17 See J. Feiler and P. Vajda, "Energy and environment," in Leal-Arcas, R. and Wouters, J. *Research Handbook on EU Energy Law and Policy*, Edward Elgar, 2017, pp. 432–449.

18 Janardhanan, N. "The new energy order and its impact on the energy policy of Asian countries," *Asian Energy Institute Newsletter*, Issue 11, p. 4.

19 Sawin, J. et al., "Renewables Global Status Report," REN21, 2012; McHarg, A. et al. (eds), *Property and the Law in Energy and Natural Resources*, Oxford University Press, 2010; Barton, B. et al. (eds), *Regulating Energy and Natural Resources*, Oxford University Press, 2006.

20 See generally Leal-Arcas, R., Grasso, C. and Alemany Rios, J. *Energy Security, Trade and the EU: Regional and International Perspectives*, Edward Elgar, 2016.

21 Rodrik, D. (2011) *The Globalization Paradox: Democracy and the Future of the World Economy*, New York, W.W. Norton & Co.

22 See for example ICTSD, "Linking Trade, Climate Change and Energy," ICTSD, Geneva, Switzerland, 2006.

23 Pauwelyn, J. "Global Challenges at the Intersection of Trade, Energy and the Environment: An Introduction," in Pauwelyn, J. (ed.), *Global Challenges at the Intersection of Trade, Energy and the Environment*, Geneva, Switzerland, Centre for Trade and Economic Integration, 2010, p. 3.

24 Gault, J. "A World of Introduction from the Energy Industry Perspective," in Pauwelyn, J. (ed.), *Global Challenges at the Intersection of Trade, Energy and the Environment*, Geneva, Switzerland, Centre for Trade and Economic Integration, 2010, p. 9.

25 WTO, "Members and Observers," available at www.wto.org/english/thewto_e/whatis_e/tif_e/org6_e.htm.

26 Idem.

27 These countries are Iraq, Iran, Sudan, Azerbaijan, Equatorial Guinea, Algeria, and Lybia.

28 For further details, see Selivanova, Y. (ed.), *Regulation of Energy in International Trade Law: WTO, NAFTA and Energy Charter*, Kluwer, 2011; Shih, W. "Energy Security, GATT/WTO, and Regional Agreements," *Nat. Res. J.*, Vol. 49, 2009, p. 433.

29 Marceau, G. "The WTO in the Emerging Energy Governance Debate," *Global Trade and Customs Journal*, Vol. 5, Issue 3, 2010.

30 On this point, see the views of Cottier, T. et al. "Energy in WTO law and policy," in Cottier, T. and Delimatsis, P. (eds), *The Prospects of International Trade Regulation: From Fragmentation to Coherence*, New York, Cambridge University Press, 2011, pp. 211–244 (arguing that, since the regulation of energy in international law is highly fragmented and largely incoherent, pertinent issues should be addressed by a future Framework Agreement on Energy in the context of WTO law).

31 WTO, "Ministerial Conference Approves Russia's WTO Membership," 16 December 2011, available at www.wto.org/english/news_e/news_e.htm; see also "Russia to Join WTO Alone," *The Moscow Times*, 16 April 2010, available at www.themoscowtimes.com/business/article/russia-to-join-wto-alone/404038.html.

32 International Centre for Trade and Sustainable Development, "Russia Update," *Bridges Monthly*, Vol. 13, No. 2, June 2009, available at http://ictsd.net/i/news/bridges/48583/.

33 For an analysis on the governance of global energy, see the special issue "Global Energy Governance," *Global Policy*, Vol. 2, Issue Supplement s1, September 2011, pp. 1–154; Goldthau, A., J.M. Witte (eds), *Global Energy Governance: The New Rules of the Game*, Washington, DC, Brookings Institution Press, 2010; Leal-Arcas, R. et al., *International Energy Governance: Selected Legal Issues*, Edward Elgar, 2014.

34 See generally Leal-Arcas, R. *Climate Change and International Trade*, Edward Elgar, 2013.

35 Cinnamon Carlarne argues that the WTO should adopt an interpretative clause that provides clear principles for defining the relationship between the WTO and multilateral environmental agreements. See Carlarne, C. "The Kyoto Protocol and the WTO: Reconciling Tensions Between Free Trade and Environmental Objectives," *Colorado Journal of International Environmental Law and Policy*, Vol. 17, Issue 1, pp. 45–88, 2006. See also Charnovitz, S. "A New WTO Paradigm for Trade and the Environment," *Singapore Year Book of International Law*, Vol. 11, pp. 15–40, 2007; Gentile, D. "International Trade and the Environment: What is the Role of the WTO?" *Fordham Environmental Law Review*, Vol. 20, pp. 197–232, 2009; McNamee, D. "Climate Change, the Kyoto Protocol, and the World Trade Organization: Challenges and Conflicts," *Sustainable Development Law and Policy*, Vol. 6, Winter 2006, pp. 41–44.

36 See Green, A. "Climate Change, Regulatory Policy and the WTO: How Constraining are Trade Rules?" *Journal of International Economic Law*, Vol. 8, Issue 1, pp. 143–189, 2005.

37 See for instance Messerlin, P. "Climate and Trade Policies: From Mutual Destruction to Mutual Support," *World Trade Review*, Vol. 11, Issue 1, pp. 53–80, 2012.

38 John Jackson has studied the link between world trade rules and environmental policies in Jackson, J. "World Trade Rules and Environmental Policies: Congruence or Conflict?" in Bhagwati, J. and Hirsch, M. (eds), *The Uruguay Round and Beyond: Essays in Honor of Arthur Dunkel*, University of Michigan Press, 1999, pp. 414–448.

39 Ghosh and Yamarik have studied the impact of RTAs on the environment. They found that membership in an RTA reduces the amount of environmental damage by increasing the volume of trade and raising per capita income. They did not, however, find that RTAs directly impact the environment. These results suggest that recent surge of regional trading arrangements will not increase the amount of pollution, but in fact may help the environment. See Ghosh, S. & Yamarik, S. "Do Regional Trading Arrangements Harm the Environment? An Analysis of 162 Countries in 1990," *Applied Econometrics and International Development*, Vol. 6, No. 2, 2006.

40 R. Leal-Arcas "Proliferation of Regional Trade Agreements: Complementing or Supplanting Multilateralism?" *Chicago Journal of International Law*, Vol. 11, No. 2, pp. 597–629, 2011.

41 Clara Brandi, "Trade elements in countries' climate contributions under the Paris Agreement," International Centre for Trade and Sustainable Development, Issue Paper, 2017, p. 12.

42 O.J. L 127/6, 14 May 2011, signed on 6 October 2010; in force as of 13 December 2015.

43 O.J. L 354/3, 21 December 2012, signed on 26 June 2012. Provisionally applied from 1 March 2013 between the EU and Peru; provisionally applied (with the exception of Articles 2, 202(1), 291 and 292) from 1 August 2013 between the EU and Colombia.

44 O.J. L 161/3, 29 May 2014, signed on 26 June 2014, in force as of 1 September 2017.

45 Not yet in force, signed on 30 October 2016.

46 Per a European Commission press release dated 6 July 2017, Japan and the EU have only reached so far an "*agreement in principle*" which covers various but not all the aspects under negotiation (http://trade.ec.europa.eu/doclib/press/index.cfm?id=1686; accessed on 17 August 2017.)

47 Korea: Chapter Thirteen, "*Trade and sustainable development*," Article 13.5; Ukraine: Chapter 13, "*Trade and sustainable development*," Article 292; CETA: Chapter twenty-four, "*Trade and environment*," Article 24.4; JEEPA: Chapter twenty-four, "*Trade and sustainable development*," Article 4.
48 This "constant basis" has specifically been taken from Article 13.5 of the EU–South Korea FTA.
49 There are, however, minor variations on the wording. For instance, JEEPA refers to "*related practices*"; and in CETA, each party implements multilateral environmental agreements "*in its whole territory*."
50 The wording of the first paragraph, which has not been reproduced here, contains the following relevant verbs: "1. The Parties *recognise* the value . . . *stress the need to enhance* the mutual supportiveness [. . . and . . .] shall *dialogue and cooperate* as appropriate. . . ."
51 Compare this wording of paragraph 4 of Article 270 of the EU–Colombia–Peru TA with the wording of the fourth paragraph of Article 24.4 CETA: "4. The Parties *acknowledge* their right to use Article 28.3 (General exceptions) in relation to environmental measures, including those taken pursuant to multilateral environmental agreements to which they are party."
52 Only one of the instruments listed in paragraph 2 may be characterized as a climate-change instrument. Beyond this, in the EU–Colombia–Peru TA, climate change agreements are the object of a separate, notably weak provision, namely Article 275.
53 See, for instance, Article 29.2 ("*Scope*") of Chapter twenty-nine ("*Dispute settlement*") of CETA: "*Except as otherwise provided in this Agreement, this Chapter applies to any dispute concerning the interpretation or application of the provisions of this Agreement.*"
54 EU–Korea FTA, Article 13.16 ("*Dispute settlement*"): "*For any matter arising under this Chapter, the Parties shall only have recourse to the procedures provided for in Articles 13.14 and 13.15.*" (See similarly Article 300.7 of the Agreement with Ukraine, and the first paragraph of Article 24.16 of CETA.)
55 See, as examples, Article 13.15 ("*Panel of experts*") of the EU–Korea FTA; Article 16 of the "*Trade and sustainable development*" chapter of JEEPA.
56 Respectively, Article 24.15.11 of CETA, Article 285.4 of the EU–Colombia–Peru TA, and Article 17.6 of the "*Trade and sustainable development*" chapter of JEEPA.
57 Not yet in force; not yet signed by the EU or its Member States. See Opinion 2/15 [Opinion of the Court (Full Court) of 16 May 2017, ECLI:EU:C:2017:376]: "*The Free Trade Agreement between the European Union and the Republic of Singapore falls within the exclusive competence of the European Union, with the exception of the following provisions, which fall within a competence shared between the European Union and the Member States: . . ..*"
58 Article 13.6.2 of the EU–Singapore FTA.
59 This third paragraph resembles the third paragraph of the corresponding provisions of JEEPA and of the EU–Korea FTA.
60 See Article 13.16 ("*Government Consultations*"): "1. In case of disagreement on any matter arising under this Chapter, the Parties *shall only* have recourse to the procedures provided for in Article 13.16 (Government Consultations) and Article 13.17 (Panel of Experts). *Chapter Fifteen (Dispute Settlement)* and Chapter Sixteen (Mediation Mechanism) *do not apply to this Chapter.*"
61 Following a meeting report of the "*Transatlantic Trade & Investment Partnership Advisory Group*" dated 9 March 2017, the negotiations are "*frozen*" (p. 3) [http://trade.ec.europa.eu/doclib/docs/2017/april/tradoc_155484.pdf, accessed on 16 August 2017].
62 http://trade.ec.europa.eu/doclib/docs/2015/november/tradoc_153923.pdf.
63 http://trade.ec.europa.eu/doclib/docs/2016/july/tradoc_154800.pdf, accessed on 18 August 2017.

64 Further paragraphs in the draft provisions resort to language which has mostly already been discussed in previous pages of this chapter.

65 Available at http://trade.ec.europa.eu/doclib/docs/2016/july/tradoc_154800.pdf.

66 Regarding the protection of the ozone layer ad measures related to hydrofluorocarbons, see text available at http://trade.ec.europa.eu/doclib/docs/2016/july/tradoc_154800.pdf.

67 New York Times, "Trump Will Withdraw U.S. From Paris Climate Agreement," 1 June 2017 (www.nytimes.com/2017/06/01/climate/trump-paris-climate-agreement.html?mcubz=1; accessed on 22 August 2017).

Index

Taylor & Francis Group
an **informa** business

Taylor & Francis eBooks

www.taylorfrancis.com

A single destination for eBooks from Taylor & Francis with increased functionality and an improved user experience to meet the needs of our customers.

90,000+ eBooks of award-winning academic content in Humanities, Social Science, Science, Technology, Engineering, and Medical written by a global network of editors and authors.

TAYLOR & FRANCIS EBOOKS OFFERS:

- A streamlined experience for our library customers
- A single point of discovery for all of our eBook content
- Improved search and discovery of content at both book and chapter level

REQUEST A FREE TRIAL

support@taylorfrancis.com

Routledge
Taylor & Francis Group

CRC Press
Taylor & Francis Group

For Product Safety Concerns and Information please contact our
EU representative GPSR@taylorandfrancis.com Taylor & Francis
Verlag GmbH, Kaufingerstraße 24, 80331 München, Germany